SIGNIFICANT CHANGES TO THE

INTERNATIONAL FIRE CODE®

2018 EDITION

SIGNIFICANT CHANGES TO THE INTERNATIONAL FIRE CODE®

2018 EDITION

Kevin H. Scott, Author

Additional contributions by Robert A. Neale

International Code Council

ICC Staff :

Executive Vice President and Director of Business Development: Mark A. Johnson

Senior Vice President, Business and Product Development: Hamid Naderi

Vice President and Technical Director of Products and Services: Doug Thornburg

Senior Marketing Specialist: Dianna Hallmark

ISBN: 978-1-60983-712-9

Cover Design: Lisa Triska
Project Head: Steve Van Note
Publications Manager: Mary Lou Luif

INTERNATIONAL CODE COUNCIL®

Errata on various ICC publications may be available at www.iccsafe.org/errata.

First Printing: September 2017

PRINTED IN THE U.S.A.

T023580

Contents

Preface

The purpose of *Significant Changes to the International Fire Code®, 2018 Edition*, is to familiarize fire officials, building officials, plans examiners, fire inspectors, design professionals and others with many of the important changes in the 2018 *International Fire Code®* (IFC®). This publication is designed to assist code users in identifying the specific code changes that have occurred and, more important, in understanding the reasons behind the changes. It is also a valuable resource for jurisdictions in order to help them explain the significance and impact of the changes as they go through their code adoption process.

Only a portion of the total number of code changes to the IFC are discussed in this book. The changes selected were identified for a number of reasons, including their frequency of application, special significance or change in application. However, the importance of the changes not included is not to be diminished. Further information on all code changes can be found in the Complete Revision History, available from the International Code Council® (ICC®) in 2018, through the online store at http://shop.iccsafe.org. This resource collection provides the published documentation for each successful code change contained in the 2018 IFC since the 2015 edition.

Significant Changes to the International Fire Code, 2018 Edition, is arranged to follow the general layout of the IFC, including code sections and section number format. The table of contents, in addition to providing guidance in the use of this publication, allows for a quick identification of those significant code changes that occur in the 2018 IFC.

Throughout the book, each change is accompanied by a photograph or an illustration to assist in and enhance the reader's understanding of the specific change. A summary and a discussion of the significance of the change are also provided. Each code change is identified by type, be it an addition, modification, clarification or deletion.

The code change itself is presented in a legislative format similar to the style utilized for code change proposals. Deleted code language is shown with a strikethrough, and new code text is indicated by underlining. As a result, the actual 2018 code language is provided as well as a comparison with the 2015 language, so the user can easily determine changes to the specific code text.

As with any code change text, *Significant Changes to the International Fire Code,* 2018 Edition, is best used as a companion to the 2018 IFC. Because only a limited discussion of each change is provided, the reader should reference the code itself in order to gain a more comprehensive understanding of the code change and its application.

The commentary and opinions set forth in this text are those of the authors and do not necessarily represent the official position of ICC. In addition, they may not represent the views of any enforcing agency because such agencies have the sole authority to render interpretations of the IFC. In many cases, the explanatory material is derived from the reasoning expressed by code change proponents.

Comments concerning this publication are encouraged and may be directed to ICC at significantchanges@iccsafe.org.

About the *International Fire Code*

Fire code officials, fire inspectors, building officials, design professionals, contractors and others involved in the field of fire safety recognize the need for a modern, up-to-date fire code. The *International Fire Code* (IFC), 2018 Edition, is intended to meet these needs through model code regulations that safeguard the public health and safety in all communities, large and small. The IFC is kept up to date through ICC's open code development process. The provisions of the 2015 edition, along with those code changes approved through 2016, make up the 2018 edition.

One in a family of International Codes® published by ICC, the IFC is a model code that establishes minimum fire safety requirements for new and existing buildings, facilities, storage and processes. It addresses fire prevention, fire protection, life safety and safe storage and use of hazardous materials. The IFC provides a total approach of controlling hazards in all buildings and sites, regardless of the hazard being indoors or outdoors.

The IFC is a design document. For example, before a building is constructed, the site must be provided with an adequate water supply for fire-fighting operations and a means of building access for emergency responders in the event of a medical emergency, fire or natural or technological disaster. Depending on the building's occupancy and uses, the IFC regulates the various hazards that may be housed within the building, including refrigeration systems, application of flammable finishes, fueling of motor vehicles, high-piled combustible storage and the storage and use of hazardous materials. The IFC sets forth minimum requirements for these and other hazards and contains requirements for maintaining the life safety of building occupants, the protection of emergency responders, and to limit the damage to a building and its contents as the result of a fire, explosion or unauthorized hazardous material discharge and electrical systems. The IFC is available for adoption and use by jurisdictions internationally. Its use within a governmental jurisdiction is intended to be accomplished through adoption by reference, in accordance with proceedings establishing the jurisdiction's laws.

Acknowledgements

Significant Changes to the International Fire Code, 2018 Edition, is the result of a collaborative effort, and the author is grateful for the assistance and contributions by the following talented staff of ICC Business and Product Development: Hamid Naderi, P.E., C.B.O., Senior Vice President; Doug Thornburg, A.I.A., C.B.O., Vice President, Director of Education and Certification; and Jay Woodward, A.I.A, Senior Staff Architect. The author would also like to thank Bob Davidson, Davidson Code Concepts, LLC; Howard Hopper, Regulatory Services Program Coordinator with Underwriters Laboratories LLC; and Pat McLaughlin, McLaughlin & Associates for their insight and assistance with application of some of the technical requirements.

A special thank you is extended to Robert Neale, Vice President, ICC Government Relations. Rob spent many hours working on the content, sharing his knowledge, and providing peer review.

About the Author

Kevin H. Scott
President
KH Scott & Associates LLC

Kevin Scott is President of KH Scott & Associates LLC. Kevin has extensive experience in the development of fire safety, building safety and hazardous materials regulations. Kevin has actively worked for over 25 years in the development of fire code, building code and fire safety regulations at the local, state, national and international levels. Kevin previously worked as a Senior Regional Manager with the International Code Council, and before that, he was Deputy Chief for the Kern County Fire Department, California, where he worked for 30 years. He has developed and presented many seminars on a variety of technical subjects including means of egress, high-piled combustible storage, hazardous materials, and plan review and inspection practices.

Kevin was a member of the original IFC Drafting Committee that worked to create the first edition of the IFC. He served for seven years on the IFC Code Development Committee and was chairperson for the committee from 2001 to 2004. Kevin has actively participated in numerous technical committees to evaluate specific hazards and technologies, and to create regulations specific to those hazards.

Some of the more significant committees are:

- High-piled Combustible Storage Committee
- Hydrogen Gas Ad Hoc Committee
- Task Group 400
- Technical Advisory Committee on Retail Storage of Group 'A' Plastic Commodities
- Underwriters Laboratories Fire Council.

Kevin's constant work to improve fire and life safety has been recognized on many levels. His contributions have been acknowledged by various organizations when they presented him with the following awards:

- Mary Eriksen-Rattan Award in 2013—presented by the Southern California State Fire Prevention Officers' Association
- William Goss Award in 2009—presented by the California State Firefighters Association
- Fire Official of the Year Award in 2005—presented by the California Building Officials
- Robert W. Gain Award in 2003—presented by the International Fire Code Institute.

About the Contributor

Robert A. Neale

Vice President

Governement Relations: National Fire Service Activities

Rob currently serves as the International Code Council Vice President for Government Relations: National Fire Service Activities. He is responsible for strategic guidance to help local fire organizations adopt and enforce the most recent version of the model codes and build relationships among code enforcement entities.

In 2015, Rob retired as Deputy Superintendent for the United States Fire Administration National Fire Academy in Emmitsburg, Maryland. He led the development of curriculum aimed at improving the professionalism of America's fire service.

In 2012, he received the FEMA Administrator's Award for Innovation in training for his development of the popular Coffee Break Training series. Rob was awarded the International Society for Performance Improvement award for the National Fire Academy class "Evaluating Performance-Based Designs" in 2004.

He has more than 30 years of experience in Washington state municipal fire protection as a fire chief, fire marshal and fire fighter.

Rob has a master's degree from the Center for Homeland Defense and Security at the Naval Postgraduate School and a bachelor's degree in Liberal Studies from Western Washington University.

Rob has been published regularly in national fire protection trade journals, and for many years has been content developer and platform instructor on codes, standards and fire protection systems.

About the International Code Council

The International Code Council is a member-focused association. It is dedicated to developing model codes and standards used in the design, build and compliance process to construct safe, sustainable, affordable and resilient structures. Most U.S. communities and many global markets choose the International Codes. ICC Evaluation Service® (ICC-ES®) is the industry leader in performing technical evaluations for code compliance fostering safe and sustainable design and construction.

ICC Headquarters:
500 New Jersey Avenue, NW, 6th Floor,
Washington, DC 20001

Regional Offices:
Birmingham, AL • Chicago, IL • Los Angeles, CA

1-888-422-7233 (ICC-SAFE)
www.iccsafe.org

PART 1

Administration

Chapters 1 and 2

- **Chapter 1** Scope and Administration
  No changes addressed
- **Chapter 2** Definitions No changes addressed

The provisions of Chapter 1 set forth requirements for the adoption, application, enforcement, and administration of the *International Fire Code* (IFC). In addition to establishing the general scope and intent of the IFC, the chapter addresses construction, operational and maintenance inspection authority, the responsibilities and authority of the fire code official, and the adoption of referenced standards. Upon adoption by a governmental jurisdiction, Chapter 1 establishes requirements for the issuance of construction and operational permits, inspection of facilities and processes, collection of fees, issuance of stop work orders and other legal notices, and formation of a board of appeals.

Chapter 2 provides definitions for terms used throughout the IFC and the *International Building Code*® (IBC®). Codes are technical documents, and every word and term can add to or change the meaning of a provision and its application. ■

PART 2

General Safety Provisions

Chapters 3 and 4

- **Chapter 3** General Requirements
- **Chapter 4** Emergency Planning and Preparedness

A basic requirement of the *International Fire Code* (IFC) is to prevent the ignition of materials inside and outside buildings. Controlling fuels and ignition sources limits the potential for fire. Chapter 3 contains requirements for combustible waste materials, control or elimination of ignition sources, open flames, recreational fires, and the use of smoking materials. Certain equipment can also be a source of ignition, and Chapter 3 addresses the proper operation of asphalt kettles and powered industrial trucks. In occupancies such as assembly uses or covered malls, controls are specified for certain hazardous materials or displays of vehicles. Chapter 3 also contains requirements for the protection from vehicle impact to prevent the release of compressed gas, flammable liquid or hazardous materials.

Chapter 4 requires that evacuation plans be prepared, that a hazardous materials communication program be established, and that employees be trained to identify fire hazards and safely evacuate building occupants. ■

314.4

Indoor Display of Vehicles

315.3.1

Ceiling Clearance for Indoor Storage

315.1, 315.7

Outdoor Pallet Storage

403.12.3

Crowd Managers

404.2.3

Lockdown Plans

314.4

Indoor Display of Vehicles

CHANGE TYPE: Modification

CHANGE SUMMARY: This section is revised to clarify it applies to both liquid-fueled vehicles and gaseous-fueled vehicles. Additionally, it has been modified to allow the Fire Code Official the ability to determine the best method of safeguarding the vehicle regarding the battery and electrical system.

2018 CODE: 314.4 Vehicles. Liquid-fueled or ~~gas-fueled~~ gaseous-fueled vehicles, boats or other motorcraft shall not be located indoors except as follows:

1. Batteries are disconnected except where the *fire code official* requires that the batteries remain connected to maintain safety features.
2. Fuel in fuel tanks does not exceed one-quarter tank or 5 gallons (19 L) (whichever is least).
3. Fuel tanks and fill openings are closed and sealed to prevent tampering.
4. Vehicles, boats or other motorcraft equipment are not fueled or defueled within the building.

International Code Council®

The battery on vehicles located indoors needs to remain connected in situations where disconnection will result in loss of safety monitoring systems.

CHANGE SIGNIFICANCE: This section provides for the safe display of fueled vehicles indoors. The first sentence refers to liquid-fueled vehicles, which includes those using gasoline, diesel, bio-diesel, and similar fuels and gaseous-fueled vehicles and intends to apply to vehicles fueled by natural gas, propane, and more recently, hydrogen. However, the use of the word "gas," which is also the generic term used for liquid gasoline, leads to some confusion. Therefore, the term "gas-fueled" is replaced with "gaseous-fueled" vehicles.

For traditional fuels, such as gasoline or diesel fuel, this section is very clear and appropriate regarding disconnecting the vehicle battery.

However, technology has advanced and brought alternative fueled vehicles. Some of these alternative fueled vehicles have gas alarms that would be disabled if the batteries were disconnected. Item 1 has been revised to allow the Fire Code Official to determine which method of safeguarding is most appropriate for the vehicle in question. In some cases, it may be more appropriate to leave the batteries connected and the safety systems in operation. For example, many newer vehicles with alternative fuels (hydrogen, CNG, LNG, etc.) have built-in sensors to detect leaks or overpressure. Disconnecting the batteries will drop power to the sensors, disabling these safety devices.

315.3.1

Ceiling Clearance for Indoor Storage

CHANGE TYPE: Modification

CHANGE SUMMARY: Exceptions have been added which allow an increase in the height of storage along walls in sprinklered and nonsprinklered buildings.

2018 CODE: 315.3.1 Ceiling clearance. Storage shall be maintained 2 feet (610 mm) or more below the ceiling in nonsprinklered areas of buildings or not less than 18 inches (457 mm) below sprinkler head deflectors in sprinklered areas of buildings.

Exceptions:

1. The 2-foot (610 mm) ceiling clearance is not required for storage along walls in nonsprinklered areas of buildings.
2. The 18-inch (457 mm) ceiling clearance is not required for storage along walls in areas of buildings equipped with an *automatic sprinkler system* in accordance with Section 903.3.1.1, 903.3.1.2 or 903.3.1.3.

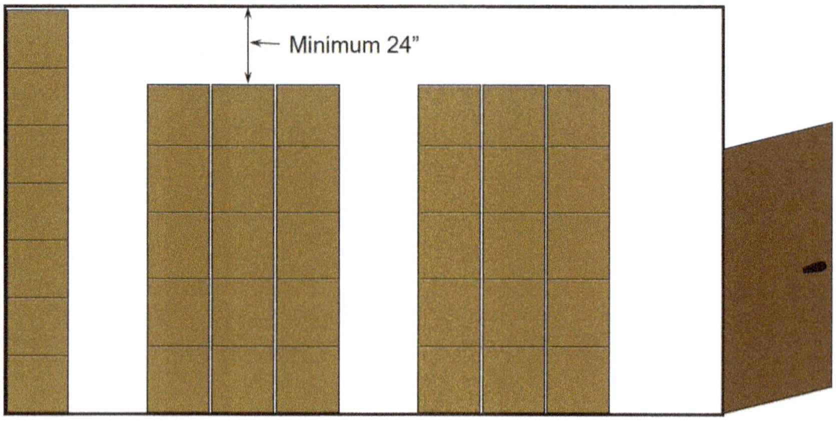

Storage in a nonsprinklered room can extend to the ceiling along the walls, but must maintain 24 inches clear to the ceiling in the remainder of the room.

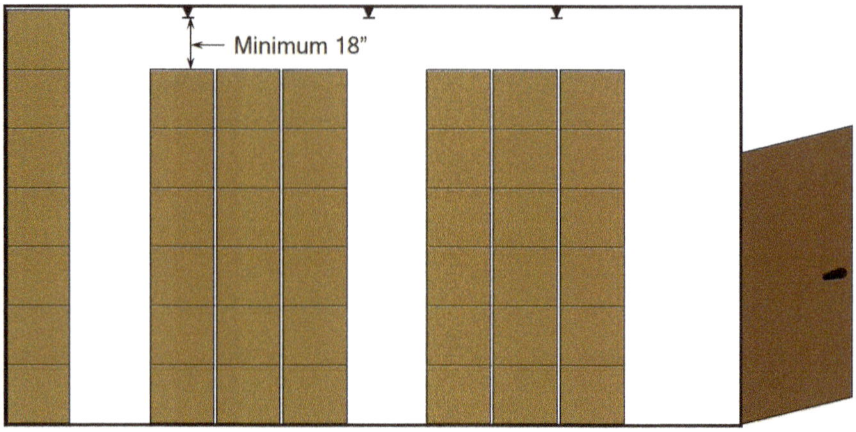

Storage in a room with sprinklers can extend to the ceiling along the walls, but must maintain 18 inches clear below the fire sprinkler deflectors in the remainder of the room.

CHANGE SIGNIFICANCE: The limitation of storage heights in buildings addresses two conditions; a sprinklered space and a nonsprinklered space. In the sprinklered space, the top of storage is limited to 18 inches below the sprinkler deflector, while in the nonsprinklered space storage is limited to 24 inches below the ceiling.

In a nonsprinklered space, the height limit provides the ability for fire fighters to direct hose streams over stored materials to extinguish fires deeper within the room. However, the storage of materials along the wall does not impede this ability. The hose streams do not need to go over the top of the storage if the storage is along the wall.

Additionally, the installation of wall-mounted cabinets is not prohibited by the section, and could extend to the ceiling. This change recognizes that the arrangement of the storage along the wall should not be regulated differently whether it is in a cabinet or on a shelf.

In a sprinklered building, NFPA 13 Section 8.6.6.2.1 allows the top of storage on shelving on a wall to extend above the plane at 18 inches below the deflector. NFPA 13, Section 3.9.2.6 indicates that shelving, by definition, is 30 inches or less in depth. Based on these two sections, NFPA 13 stipulates that storage along the wall on shelves with a depth of 30 inches is not limited to the storage height of 18 inches below the deflector. The new exception in the IFC correlates with this allowance.

International Code Council®

NFPA 13 allows storage directly along the wall to encroach into the 18-inch separation below the sprinkler deflector.

It is important to note that this allowance is located in Chapter 3 and is a general requirement applicable to general storage arrangements. These exceptions are not located in Chapter 32 for high-piled combustible storage, nor is this section referenced by Chapter 32. As indicated in Section 102.10 of the IFC, when there is a difference between a general requirement and specific requirement, the specific requirement will apply. Applying this concept, Section 315.3.1 will apply to general storage, but not to high-piled storage arrangements.

315.1, 315.7, 105.6.29

Outdoor Pallet Storage

CHANGE TYPE: Addition

CHANGE SUMMARY: Requirements are added to the code for height limitation and separation to buildings and property lines for the outdoor storage of idle pallets constructed of wood or plastic. See also Significant Change to Section 2810 for pallet storage at pallet recycling and manufacturing facilities.

2018 CODE:

SECTION 315
GENERAL STORAGE

315.1 General. Storage shall be in accordance with Sections 315.2 through 315.6. Outdoor pallet storage shall be in accordance with Sections 315.2 and 315.7.

> **Exception:** Wood and wood composite pallets stored outdoors at pallet manufacturing and recycling facilities and complying with Section 2810.

315.7 Outdoor pallet storage. Pallets stored outdoors shall comply with Sections 315.7 through 315.7.7. Pallets stored within a building shall be protected in accordance with Chapter 32.

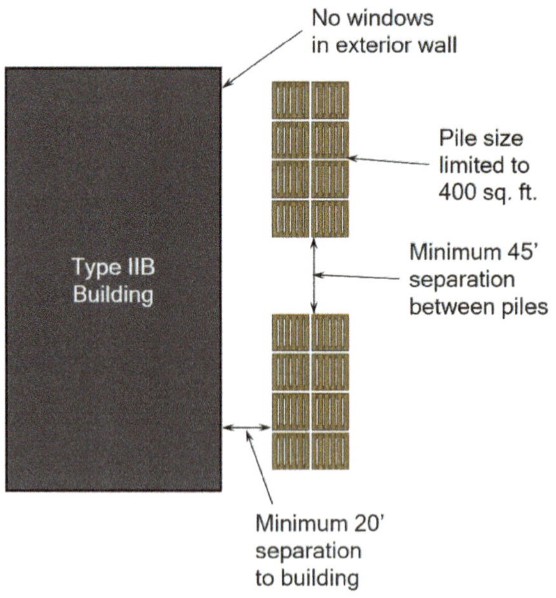

The dimensions shown are based on the requirements for more than 200 pallets in a pile. Note, if there are any windows in the exterior building wall, the separation distance changes from 20 feet to 90 feet.

315.7.1 Storage beneath overhead projections from buildings. Where buildings are equipped throughout with an *automatic sprinkler system*, the outdoor storage of pallets under eaves, canopies or other projections or overhangs are prohibited except where automatic sprinklers are installed under such eaves, canopies or other projections or overhangs.

315.7.2 Distance to lot line. Pallet storage shall not be located within 10 feet (3048 mm) of a *lot line*.

315.7.3 Storage height. Pallet storage shall not exceed 20 feet (6096 mm) in height.

315.7.4 Pallet pile stability and size. Pallet stacks shall be arranged to form stable piles. Individual pallet piles shall cover an area not greater than 400 square feet (37 m^2).

315.7.5 Pallet types. Pallets shall be all wood, with slatted or solid top or bottom, with metal fasteners, or shall be plastic or composite pallets, listed and labeled in accordance with UL 2335 or FM 4996. Plastic pallets shall be both solid and gridded deck, independent of the pallet manufacturing process, type of resin used in fabrication or geometry of the pallet.

315.7.6 Pile separation distances. In addition to the other requirements of this section, pallet stacks and piles shall be separated in accordance with Sections 315.7.6.1 and 315.7.6.2.

TABLE 315.7.6(1) Separation Distance Between Wood Pallet Stacks and Buildings

		WOOD PALLET SEPARATION DISTANCE (feet)		
WALL CONSTRUCTION	OPENINGS	50 PALLETS or LESS	51 to 200 PALLETS	GREATER than 200 PALLETS
Masonry	None	2	2	2
Masonry	Fire-rated glazing with open sprinklers	2	5	20
Masonry	Fire-rated glazing	10	5	20
Masonry	Plain glass with open sprinklers	10	5	20
Noncombustible	None	10	5	20
Wood with open sprinklers	NA	10	5	20
Wood	None	15	30	90
Any	Plain glass	15	30	90

For SI: 1 foot = 304.8 mm.

315.1, 315.7, 105.6.29 continues

315.1, 315.7, 105.6.29 continued

TABLE 315.7.6(2) Separation Distance Between Plastic Pallet Stacks and Buildings

WALL CONSTRUCTION	OPENINGS	PLASTIC PALLET SEPARATION DISTANCE (feet)		
		50 PALLETS or LESS	51 to 200 PALLETS	GREATER than 200 PALLETS
Masonry	None	2	2	2
Masonry	Fire-rated glazing with open sprinklers	10	20	50
Masonry	Fire-rated glazing	15	40	100
Masonry	Plain glass with open sprinklers	15	40	100
Noncombustible	None	15	40	100
Wood with open sprinklers	NA	15	40	100
Wood	None	30	80	150
Any	Plain glass	30	80	150

For SI: 1 foot = 304.8 mm.

TABLE 315.7.6(3) Separation from Other Pallet Piles and On-site Storage (wood pallets)

	WOOD PALLET SEPARATION DISTANCE (feet)		
	50 PALLETS or LESS	51 to 200 PALLETS	GREATER than 200 PALLETS
Between pallet piles	7.5	15	45
Other on-site storage	7.5	15	45

For SI: 1 foot = 304.8 mm.

TABLE 315.7.6(4) Separation from Other Pallet Piles and On-site Storage (plastic pallets)

	PLASTIC PALLET SEPARATION DISTANCE (feet)		
	50 PALLETS or LESS	51 to 200 PALLETS	GREATER than 200 PALLETS
Between pallet piles	15	40	75
Other on-site storage	15	40	75

For SI: 1 foot = 304.8 mm.

315.7.6.1 Building separation. Pallet stacks and piles shall be separated from buildings in accordance with Table 315.7.6(1) for wood pallets and Table 315.7.6(2) for plastic pallets.

315.7.6.2 Separation from other pallets and on-site storage. Pallets shall be separated from other pallet piles and other storage in accordance with Table 315.7.6(3) for wood pallets and Table 315.7.6(4) for plastic pallets.

315.7.7 Prohibited locations. Pallets shall not be stored underneath high-voltage transmission lines, elevated roadways or elevated railways.

SECTION 105
PERMITS

105.6.29 Miscellaneous combustible storage. An operational permit is required to store in any building or on any premises in excess of 2,500 cubic feet (71 m^3) gross volume of combustible empty packing cases, boxes, barrels or similar containers, combustible pallets, rubber tires, rubber, cork or similar combustible material.

CHANGE SIGNIFICANCE: Idle pallets create a severe fire condition and radiate a significant quantity of heat to exposures. Stacks of pallets against the exterior wall of a building have repeatedly resulted in fire spreading into the building when the fire occurs in the pallet area. These new requirements address the exposure hazard created by the storage of idle pallets.

Combustible pallets are added to the list of stored combustible materials which require a permit.

For pallets stored outdoors, storage height is limited to 20 feet in Section 315.7.3 and Tables 315.6.7(1) through 315.6.7(4) provide separation criteria. Tables 315.6.7(1) and 315.6.7(3) address wood pallets. Based on Section 315.7.5, listed plastic pallets are treated the same as wood pallets, so would also fall under Tables 315.6.7(1) and 315.6.7(3). Tables 315.6.7(2) and 315.6.7(4) are for plastic pallets and would include all plastic pallets which are not listed.

The separation distance to buildings is based on two conditions. First, the quantity of pallets; and second, the type of building construction. As the fire resistance of the building wall increases, the separation can be reduced. As the number of pallets increases, the separation increases because the heat generated will also increase. It should be noted that even for less than 50 pallets adjacent to a masonry wall with no openings, there is still a required separation distance of 2 feet.

Note that the separation distance to buildings is different for outdoor pallet storage than the distance found in the Section 2810.7. (See Significant Change to IFC Section 2810.) This section covers general pallet storage at any industrial or commercial facility, whereas Section 2810 is specific to pallet manufacturing and pallet recycling facilities.

Section 315.7.4 limits the pile sizes to 400 square feet each. Tables 315.7.6(3) and 315.6.7(4) provide the separation distance between piles. For example, piles of over 200 pallets are separated into piles no more than 20 feet in height, each pile does not exceed 400 square feet, and multiple piles are separated by 45 feet, based on Table 315.7.6(3).

403.12.3, 403.12.3.1

Crowd Managers

International Code Council®

Crowd managers are required to be trained. Training is available via the internet and a link is available through the ICC Preferred Provider Program at www.crowdmanagers.com.

CHANGE TYPE: Modification

CHANGE SUMMARY: The threshold for crowd managers dropped from 1,000 to 500 people for certain events.

2018 CODE: **403.12.3 Crowd managers.** ~~**for gatherings exceeding 1,000 people.**~~ Where facilities or events involve a gathering of more than ~~1,000~~ 500 people, crowd managers shall be provided in accordance with Sections 403.12.3.1 through 403.12.3.3.

403.12.3.1 Number of crowd managers. ~~The minimum number of~~ Not fewer than two trained crowd managers ~~shall be established at a ratio of~~ , and not fewer one trained crowd manager for ~~every~~ each 250 persons or portion thereof, shall be provided for the gathering.

> ~~**Exception:** Where approved by the *fire code official*, the number of crowd managers shall be permitted to be reduced where the facility is equipped throughout with an approved *automatic sprinkler system* or based upon the nature of the event.~~

Exceptions:

1. Outdoor events with less than 1,000 persons in attendance shall not require crowd managers.
2. Assembly occupancies used exclusively for religious worship with an occupant load not exceeding 1,000 shall not require crowd managers.
3. The number of crowd managers shall be reduced where, in the opinion of the *fire code official*, the fire protection provided by the facility and the nature of the event warrant a reduction.

CHANGE SIGNIFICANCE: This section is revised to accomplish two main changes. First, the threshold where crowd managers are required moves from 1,000 to 500 persons. However, two new exceptions allow for occupant load to stay at 1,000 for places of religious worship and outdoor events. Smaller venues sometimes place the public at greater risk than larger ones for many reasons, including the fact that larger facilities have greater requirements for other fire protection features. Many times, the event itself creates additional hazards that impact smaller audiences.

The new Exception 1 for outdoor events with less than 1,000 is intended to recognize the fact that outdoor events are perceived to be less dangerous because egress isn't restricted as it is in a building. Exception 2 for places of worship with occupant loads up to 1,000 recognizes the fact that people who are in facilities normally have a greater awareness of their surroundings and are more familiar with egress routes because they attend the facility on a more regular basis than those at performances, who tend to be more transient.

Second, the section is modified to prescribe "trained" crowd managers. This concept is no different than requiring that the designated crowd manager is qualified, especially since each crowd manager could be responsible for the safety of 250 people.

The issue of providing crowd managers does not require a facility to hire any additional personnel; rather, it requires that a requisite number of their staff receive approved training in fire prevention, evacuation methods, and other duties. This training is currently provided at minimal cost across the nation. Training is available via the internet and can be found through the ICC Preferred Provider Program, or at www.crowdmanager.com.

404.2.3, 404.2.3.1, 404.2.3.2, 404.2.3.3

Lockdown Plans

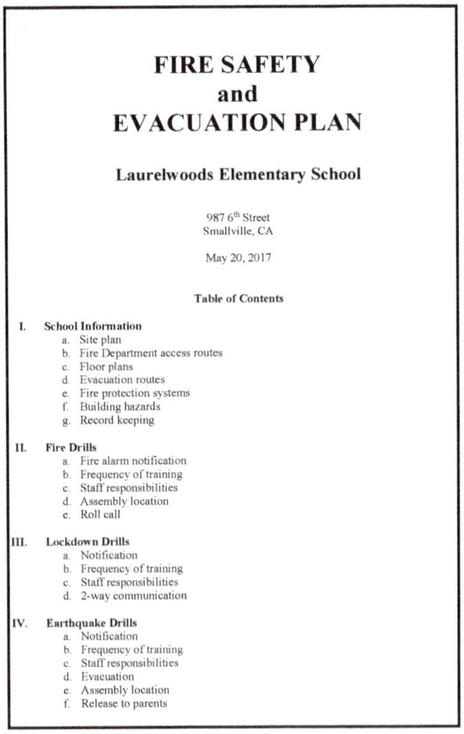

FIRE SAFETY
and
EVACUATION PLAN

Laurelwoods Elementary School

987 6th Street
Smallville, CA

May 20, 2017

Table of Contents

I. **School Information**
   a. Site plan
   b. Fire Department access routes
   c. Floor plans
   d. Evacuation routes
   e. Fire protection systems
   f. Building hazards
   g. Record keeping

II. **Fire Drills**
   a. Fire alarm notification
   b. Frequency of training
   c. Staff responsibilities
   d. Assembly location
   e. Roll call

III. **Lockdown Drills**
   a. Notification
   b. Frequency of training
   c. Staff responsibilities
   d. 2-way communication

IV. **Earthquake Drills**
   a. Notification
   b. Frequency of training
   c. Staff responsibilities
   d. Evacuation
   e. Assembly location
   f. Release to parents

Table of contents for a sample fire safety and evacuation plan.

CHANGE TYPE: Addition

CHANGE SUMMARY: Updates and prescribes details for facility lockdown plans.

2018 CODE: 404.2.3 Lockdown plans. ~~Where facilities develop a lockdown plan, it shall be in accordance with Sections 404.2.3.1 through 404.2.3.3.~~ Lockdown plans shall only be permitted where such plans are approved by the *fire code official* and are in compliance with Sections 404.2.3.1 and 404.2.3.2.

404.2.3.1 Lockdown plan contents. Lockdown plans shall ~~*be approved by the fire code official and shall*~~ include the following:

~~1. Initiation. The plan shall include instructions for reporting an emergency that requires a lockdown.~~

~~2. Accountability. The plan shall include accountability procedures for staff to report the presence or absence of occupants.~~

~~3. Recall. The plan shall include a prearranged signal for returning to normal activity.~~

~~4. Communication and coordination. The plan shall include an approved means of two-way communication between a central location and each secured area.~~

1. Identification of individuals authorized to issue a lockdown order.

2. Security measures used during normal operations, when the building is occupied, that could adversely affect egress or fire department operations.

3. A description of identified emergency and security threats addressed by the plan, including specific lockdown procedures to be implemented for each threat condition.

4. Means and methods of initiating a lockdown plan for each threat, including:

   4.1. The means of notifying occupants of a lockdown event, which shall be distinct from the fire alarm signal.

   4.2. Identification of each door or other access point that will be secured,

   4.3. A description of the means or methods used to secure doors and other access points.

   4.4. A description of how locking means and methods are in compliance with the requirements of this code for egress and accessibility.

5. Procedures for reporting to the fire department any lockdown condition affecting egress or fire department operations.

6. Procedures for determining and reporting the presence or absence of occupants to emergency response agencies during a lockdown.

7. Means for providing two-way communication between a central location and each area subject to being secured during a lockdown.

8. Identification of the prearranged signal for terminating the lockdown.

9. Identification of individuals authorized to issue a lockdown termination order.

10. Procedures for unlocking doors and verifying that the means of egress has been returned to normal operations upon termination of the lockdown.

11. Training procedures and frequency of lockdown plan drills.

404.2.3.2 ~~Training frequency~~ Drills. ~~The training frequency shall be included in the lockdown plan. The lockdown drills shall not substitute for any of the fire and evacuation drills required in Section 405.2.~~ Lockdown plan drills shall be conducted in accordance with the approved plan. Such drills shall not be substituted for fire and evacuation drills required by Section 405.2.

~~**404.2.3.3 Lockdown notification.** The method of notifying building occupants of a lockdown shall be included in the plan. The method of notification shall be separate and distinct from the fire alarm signal.~~

CHANGE SIGNIFICANCE: Given the recent increase in school and workplace violence, this proposal is intended to address the various security means and methods that have been proposed by parents, teachers and, in some cases, state legislators. While the well-intended proposals address criminal activities, many of these proposals have unintended consequences in their effect on emergency egress and fire and life safety systems.

This change adds specific topics that need to be addressed in a lockdown plan. One specific issue is the need to identify the organizational structure in charge of making decisions during a lockdown. Clear authorities and communications must be addressed for a plan to succeed.

When developing a lockdown plan, it is important to consider that all threats are not equal. Procedures suitable for an unwanted chemical release will be substantially different from those during an active shooter incident. The revisions in these sections allow for flexibility in the plan, but also require that the plan is approved by the fire code official.

Item 4.3 requires a description of the means and methods of locking doors and securing areas. Part of the review by the fire code official would evaluate the method of securing doors. There are many devices and mechanisms available, but some also present a detrimental impact on the means of egress. The lockdown plan must specify all methods used for security, and demonstrate compliance with IFC Chapter 10 requirements. Item 10 also requires the establishment of a process to ensure that after the lockdown is cancelled, that all doors and egress paths are again available and unhindered egress is provided.

PART **3**

Building and Equipment Design Features

Chapters 5 through 12

- **Chapter 5** **Fire Service Features**
- **Chapter 6** **Building Services and Systems**
- **Chapter 7** **Fire and Smoke Protection Features**
  No changes addressed
- **Chapter 8** **Interior Finish, Decorative Materials and Furnishings**
- **Chapter 9** **Fire Protection Systems**
- **Chapter 10** **Means of Egress**
- **Chapter 11** **Construction Requirements for Existing Buildings**
- **Chapter 12** **Energy Systems**
- **Chapters 13 through 19** **(Reserved)**

International Fire Code (IFC) Part III contains requirements that provide fire fighters with a means of accessing a building and establishing a fire protection water supply for that building. Chapter 5 contains requirements for address numbers on buildings and provisions for fire department access roadways. Chapter 6 contains requirements for building systems, such as elevators, standby and emergency power systems (EPS), stationary battery systems and refrigeration systems. When specified by the *International Building Code* (IBC), buildings constructed using fire-resistant materials must be properly maintained to ensure the specified fire-resistance ratings are maintained. Chapter 7 specifies the requirements for maintenance of fire-resistance-rated construction.

Interior finish and decorative materials or furnishings offer fuel contribution and surfaces through which a fire can spread and transport heat and smoke to other parts of a room or to other rooms. Chapter 8 contains requirements for regulating wall and ceiling finishes, decorative materials and furnishings.

Fire protection systems are required in accordance with Chapter 9. Chapter 9 specifies the requirements for automatic sprinkler systems, alternative fire-extinguishing systems, fire alarm and detection systems, standpipes, portable fire extinguishers and other fire appliances, emergency alarm systems, smoke and heat vents, smoke control systems, carbon monoxide alarms, and gas detection systems. For materials that can have a detonation or deflagration hazard, Chapter 9 specifies the requirements for explosion control systems.

Merriam-Webster's *New Collegiate Dictionary* defines egress as "a place or means of going out." In the event of a fire or an emergency that requires the occupants to safely exit a building, Chapter 10 establishes the minimum requirements for means of egress from buildings.

Chapter 11 sets forth retroactive construction requirements for existing buildings. These provisions establish minimum fire-resistance rating requirements for shafts as well as minimum means of egress requirements in existing buildings. It also establishes retroactive requirements for the installation of automatic sprinkler systems and fire alarm systems in existing buildings or occupancies.

Chapter 12 is added to the IFC to address the proliferation of energy storage and energy generation systems. This chapter includes what previously was in Section 604 Emergency and Standby Power, Section 605.11 Solar Photovoltaic Power Systems, and Section 608 Stationary Storage Battery Systems and expands these provisions. This chapter includes other battery technologies, fuel cell power systems, and energy storage systems, such as capacitor energy storage. ■

510

Emergency Responder Radio Coverage

603.1, 603.3

Fuel-fired Appliances

605.13, 605.16, 605.17

Refrigerants with Lower Flammability Hazards

608.3

Nonmetallic Cooking Oil Storage Tanks

807.1, 807.2, 807.5

Combustible Decorative Materials

807.4

Artificial Decorative Vegetation

901.4.6.1, 901.4.6.2, 901.4.6.3, 901.4.6.4

Fire Pump and Fire Sprinkler Riser Rooms

901.6.2

Integrated Fire Protection System Testing

901.8.2

Removal of Occupant-use Hose Lines

903.2.1

Sprinklers in Group A Occupancies

903.2.3

Sprinklers in Group E Occupancies

903.3.1.1.2

Sprinklers in Bathrooms in Group R Occupancies

903.3.1.2.1

Sprinklers Beneath Balconies

903.3.1.2.3

Protection of Attics in Group R Occupancies

903.3.3

Sprinkler Obstructions

904.12

Commercial Cooking Operations

904.13

Domestic Cooking in Institutional Occupancies

904.14, 901.6.1

Aerosol Fire-extinguishing Systems

905.3.1

Class III Standpipes

905.4

Class I Standpipe Hose Connections

905.11

Locking Caps on Standpipe Outlets

906.1

Portable Fire Extinguishers

907.1.2

Fire Alarm Construction Documents

907.2.1

Fire Alarms in Group A Occupancies

907.2.10 (DELETED)

Group R-4 Fire Alarm System

907.5.2.2.4

Emergency Voice/Alarm Communication System Captions

910.5

Maintenance of Smoke and Heat Removal Equipment

916

Gas Detection Systems

TABLE 1004.5, 1004.8

Occupant Load Calculation in Business Use Areas

1006.2.1

Spaces with One Exit or Exit Access Doorway

1006.2.2.6, 1006.2.1, 1017.2

Groups R-3 and R-4 Protected with NFPA 13D Sprinkler System

1006.3, 1006.3.1

Exits on Adjacent Stories

1008.2.3

Illumination of the Exit Discharge

1008.3.5, 1008.2.2

Emergency Illumination in Group I-2

1009.7.2

Protection of Exterior Areas of Assisted Rescue

1010.1.1

Size of Doors

1010.1.4.4

Locking Arrangements in Educational Occupancies

1010.1.9.8

Delayed Egress

1010.1.9.9, 1010.1.9.10

Electrically Locked Egress Doors

1010.1.9.12

Locks on Stairway Doors

1010.1.10

Panic Hardware and Fire Exit Hardware

1010.3

Turnstiles

1011.6

Stairway Landings

1013.2

Floor-level Exit Signs in Group R-1

1015.6, 1015.7

Fall Arrest for Rooftop Equipment

1017.3

Common Path of Egress Travel

1023.3.1

Stairway Extension

1023.5, 1024.6

Exit Stairway and Exit Passageway Penetrations

1025.1

Luminous Egress Path Marking in Group I Occupancies

1026.4, 1026.4.1

Refuge Areas for Horizontal Exits

1029.6, 1029.6.3, 1029.7

Open-air Assembly Seating

1029.9.1

Minimum Aisle Width

1030.1

Emergency Escape and Rescue Openings

1030.1.1

Operation of Emergency Escape and Rescue Openings

1031.2.2

Locking Arrangements in Existing Educational Occupancies

1031.4

Exit Signs in Existing Buildings

1031.10

Inspection and Testing of Emergency Egress Lighting

1103.5.1

Fire Sprinklers in Existing Group A-2 Occupancies

1103.9

Carbon Monoxide Alarms in Existing Buildings

1104.16.2

Wall Openings Adjacent to Fire Escapes

1105.6.2

Fire-protection-rated Doors in Existing Group I-2

CHAPTER 12

Energy Systems

1204.5

Rapid Shutdown for Solar Photovoltaic Power Systems

1206.2

Stationary Storage Battery Systems

510

Emergency Responder Radio Coverage

CHANGE TYPE: Modification

CHANGE SUMMARY: Requirements for emergency responder radio coverage have been revised to address industry and equipment enhancements with a new reference to NFPA 1221.

2018 CODE: Due to the extent of the revisions, only the outline of Section 510 is included below. For the complete text, see the 2018 *International Fire Code.*

SECTION 510
EMERGENCY RESPONDER RADIO COVERAGE

510.1 Emergency responder radio coverage in new buildings.

510.2 Emergency responder radio coverage in existing buildings.

510.3 Permit required.

510.4 Technical requirements.

510.4.1 Emergency responder communications enhancement system signal strength.

510.4.1.1 Minimum signal strength into the building.

510.4.1.2 Minimum signal strength out of the building.

510.4.1.3 System performance.

510.4.2 System design.

510.4.2.1 Amplification systems and components.

510.4.2.2 Technical criteria.

510.4.2.3 Standby power.

510.4.2.4 Signal booster requirements.

510.4.2.5 System monitoring.

510.4.2.6 Additional frequencies and change of frequencies.

510.4.2.7 Design documents.

510.4.2.8 Radio communication antenna density.

510.5 Installation requirements.

510.5.1 Approval prior to installation.

510.5.2 Minimum qualifications of personnel.

510.5.3 Acceptance test procedure.

510.5.4 FCC compliance.

510.6 Maintenance.

510.6.1 Testing and proof of compliance.

510.6.2 Additional frequencies.

510.6.3 Nonpublic safety system.

510.6.4 Field testing.

CHANGE SIGNIFICANCE: Numerous revisions occurred in the requirements for emergency responder radio coverage systems. A new reference is added for design and installation in Section 510.4.2. NFPA 1221, "Standard for the Installation, Maintenance, and Use of Emergency Services Communications Systems" contains criteria regarding the design and testing of communication systems.

In the 2015 IFC, Section 510.5.3 requires 90-percent coverage throughout the building, while Section 510.4.2 requires 95-percent coverage. This conflict in the code has been resolved by requiring 95-percent coverage in both sections of the code.

The testing procedure is revised to now use an audibility scale to determine accuracy of the radio signal rather than measuring the signal strength. The Delivered Audio Quality must meet a minimum of 3.0 or better. This test methodology is common in the communications industry and is referred as DAQ 3. The DAQ system is a rating of 1 to 5, with 5 being the best signal transmission. (See sidebar.)

Section 510.4.1.3 has been added to allow for the use of both voice and data transmissions. Many departments are now using voice and data signals. This revision allows for both types of systems or either type of system, as long as the design criteria is met.

Revisions occurred to specify that the emergency responder radio coverage system is required to have standby power supply capable of supplying the power for 12 hours of operation at 100-percent capacity. The previous requirement only specified 24 hours without any performance criteria. Now the requirement is reduced to 12 hours, but operating at full capacity for the entire 12 hours.

The new Section 510.4.2.5 requires that the system is monitored by a fire alarm control unit. The items monitored are:

1. Loss of normal AC power supply.
2. Failure of the battery charger.
3. Malfunction of the donor antennae.
4. Failure of active RF-emitting devices.
5. Battery capacity of standby power supply falling to 70 percent of operating capacity or less.
6. Failure of critical system components.
7. The communications link between the fire alarm system and the emergency responder radio enhancement system.

A new Section 510.6.3 requires that where interference occurs as a result of other building systems, those systems must be corrected immediately or removed to eliminate the interference.

DELIVERED AUDIO QUALITY (DAQ)

DAQ 1: Transmission is unusable. Speech is present but unreadable.

DAQ 2: Speech is understandable with considerable effort. Frequent repetition is needed due to noise or distortion.

DAQ 3: Speech is understandable with slight effort. Occasional repetition is needed due to noise or distortion.

DAQ 3.5: Speech is understandable with slight effort. Occasional repetition is needed due to noise or distortion.

DAQ 4: Speech is easily understood. There is occasional noise or distortion.

DAQ 4.5: Speech is easily understood. There is infrequent noise or distortion.

DAQ 5: Speech is easily understood.

603.1, 603.3

Fuel-fired Appliances

CHANGE TYPE: Modification

CHANGE SUMMARY: Fuel oil storage allowances in Section 603 have been revised to clarify applicability to internal combustion engines, such as generators and fire pumps. Fuel oil storage is increased to 1,320 gallons if the building is sprinklered and the tank is listed to UL 142.

2018 CODE: **603.1 Installation.** The installation of nonportable ~~fuel gas~~ gas-fired appliances and systems shall comply with the *International Fuel Gas Code*. The installation of nonportable liquid fuel-fired appliances and systems shall comply with this section and the *International Mechanical Code*. The installation of all other fuel-fired appliances, other than portable internal combustion engines, oil lamps and other portable devices such as blow torches, melting pots and weed burners, shall comply with this section and the *International Mechanical Code*.

603.3.2.1 Quantity limits. One or more fuel oil storage tanks containing Class II or III *combustible liquid* shall be permitted in a building. The aggregate capacity of all ~~such~~ tanks shall not exceed the following: ~~660 gallons (2498 L).~~

~~**Exception:** The aggregate capacity limit shall be permitted to be increased to 3,000 gallons (11 356 L) of Class II or III liquid for storage in protected above-ground tanks complying with Section 5704.2.9.7, where all of the following conditions are met:~~

~~1. The entire 3,000-gallon (11 356 L) quantity shall be stored in protected aboveground tanks.~~

~~2. The 3,000-gallon (11 356 L) capacity shall be permitted to be stored in a single tank or multiple smaller tanks.~~

~~3. The tanks shall be located in a room protected by an *automatic sprinkler system* complying with Section 903.3.1.1.~~

1. 660 gallons (2498 L) in unsprinklered buildings, where stored in a tank complying with UL 80, UL 142 or UL 2085.
2. 1,320 gallons (4996 L) in buildings equipped with an *automatic sprinkler system* in accordance with Section 903.3.1.1, where stored in a tank complying with UL 142.
3. 3,000 gallons (11 356 L) where stored in protected aboveground tanks complying with UL 2085 and Section 5704.2.9.7 and the room is protected by an *automatic sprinkler system* in accordance with Section 903.3.1.1.

603.3.2.2 Restricted use and connection. Tanks installed in accordance with Section 603.3.2 shall be used only to supply fuel oil to fuel-burning equipment, generators or ~~generator equipment~~ fire pumps installed in accordance with Section 603.3.2.4. Connections between tanks and equipment supplied by such tanks shall be made using closed piping systems.

603.3.2.4 Installation. Tanks and piping systems shall be installed ~~and separated from other uses~~ in accordance with Section 915 and Chapter 13, both of the *International Mechanical Code*, as applicable.

> ~~**Exception:** Protected above-ground tanks complying with Section 5704.2.9.7 shall not be required to be separated from surrounding areas.~~

603.3.2.5 Separation. Rooms containing fuel oil tanks for internal combustion engines shall be separated from the remainder of the building by *fire barriers, horizontal assemblies*, or both, with a minimum 1-hour fire-resistance rating with 1-hour fire-protection-rated *opening protectives* constructed in accordance with the *International Building Code.*

> **Exception:** Rooms containing protected above-ground tanks complying with Section 5704.2.9.7 shall not be required to be separated from surrounding areas.

603.3.2.6 Spill containment. Tanks exceeding 55 gallon (208 L) capacity or an aggregate capacity of 1,000 gallons (3785 L) that are not provided with integral secondary containment shall be provided with spill containment sized to contain a release from the largest tank.

~~603.3.2.5~~ 603.3.2.7 Tanks in basements. *No change to text.*

CHANGE SIGNIFICANCE: The intent of this code change is to identify the difference in application of portable versus nonportable fuel-fired appliances, correlate requirements within the sections for internal combustion engines, and allow another option for storage of liquid fuels for generators and fire pumps.

Section 603.1 of the 2015 IFC confused the application of these sections when it excluded internal combustion engines. Section 603.1 was revised to provide clear direction that the nonportable, or fixed, gas-fired appliances are to comply with the *International Fuel Gas Code*® (IFGC®) while nonportable, or fixed, liquid-fired appliances are to comply with the subsections and the *International Mechanical Code*® (IMC®). Section 603.3.2.2 was also correlated to include specific fuel-fired appliances such as generators, fire pumps, and other equipment.

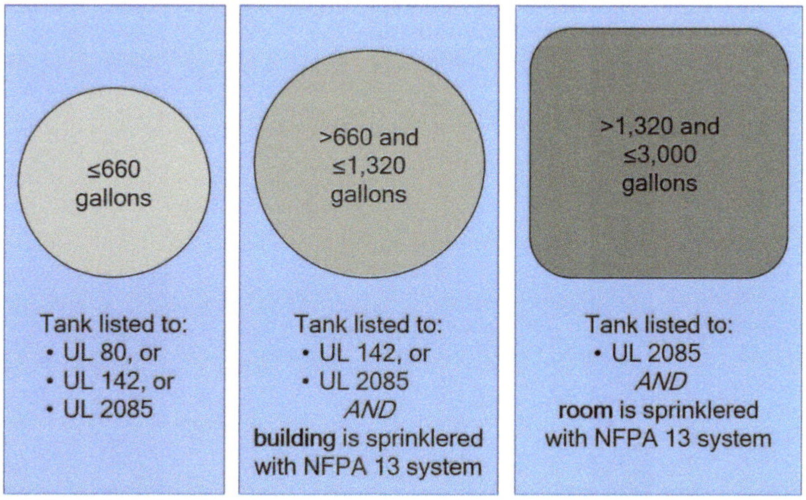

Requirements for fuel oil storage tanks inside of buildings vary based on the size of the tank.

603.1, 603.3 continues

603.1, 603.3 continued

The 2015 IFC contains two options for fuel oil storage inside buildings. These options are a limitation of a maximum of 660 gallons, or an increase to 3,000 gallons when stored in a protected above-ground tank located in a room protected with fire sprinklers. Specific provisions were added to allow a third option. The allowance for up to 660 gallons is retained, and can be accomplished in a nonsprinklered building provided a tank meets one of the standards for fuel tanks. The increase of up to 3,000 gallons where stored in a protected above-ground storage tank in a sprinkered room is also retained. A new option allows up to 1,320 gallons to be stored in a sprinklered building when it is stored in a tank listed to UL 142 or UL 2085. (See Table 603-1.)

TABLE 603-1 Maximum Capacity of Fuel Oil Based on Type of Tank and Automatic Sprinkler System Design

TANK DESIGN	NONSPRINKLERED BUILDING	FIRE SPRINKLERS PROVIDED IN THE ROOM	FIRE SPRINKLERS PROVIDED IN THE BUILDING
UL 80	660 gallons	660 gallons	660 gallons
UL 142	660 gallons	660 gallons	1,320 gallons
UL 2085	660 gallons	3,000 gallons	3,000 gallons

Additional criteria was added to specify the required construction for rooms containing fuel oil for internal combustion engines. The separation between such rooms and the remainder of the building must be at least 1-hour fire-resistance-rated construction. This requirement was in the 2015 edition of the codes, but it was often missed since the IFC referred to the IMC which referred to NFPA 37 which is where the requirement is found.

There are two important notes regarding the required separation. First, the separation requirement only applies when there is an internal combustion engine, so it would not apply to a boiler room, unless required by IBC Table 509 as an incidental use. Second, there is an exception stating the separation is not required for protected above-ground storage tanks.

Finally, Section 603.3.2.6 requires that tanks in excess of 55 gallons must be provided with spill containment.

605.13, 605.16, 605.17

Refrigerants with Lower Flammability Hazards

CHANGE TYPE: Addition

CHANGE SUMMARY: Adds requirements regarding safety concerns for lower flammability refrigerant gases.

2018 CODE: **605.13** **~~Discharge location for refrigeration machinery room ventilation~~** **Mechanical ventilation exhaust.** Exhaust from mechanical ventilation systems serving refrigeration machinery rooms containing flammable, toxic or highly toxic refrigerants, other than ammonia, capable of exceeding 25 percent of the LFL or 50 percent of the IDLH shall be equipped with *approved* treatment systems to reduce the discharge concentrations to those values or lower.

> **Exception:** Refrigeration systems containing Group A2L complying with Section 605.17.

605.16 Electrical equipment. Where *refrigerant* of Groups A2, A3, B2 and B3, as defined in the *International Mechanical Code*, are used, refrigeration machinery rooms shall conform to the Class I, Division 2 hazardous location classification requirements of NFPA 70.

> **~~Exception~~ Exceptions:**
>
> 1. Ammonia machinery rooms that are provided with ventilation in accordance with Section 1106.3 of the *International Mechanical Code.*
> 2. Machinery rooms for systems containing Group A2L *refrigerants* that are provided with ventilation in accordance with Section 605.17.

605.17 Special requirements for Group A2L refrigerant machinery rooms. Machinery rooms with systems containing Group A2L refrigerants shall comply with Sections 605.17.1 through 605.17.3.

> **Exception:** Machinery rooms conforming to the Class I, Division 2 hazardous location classification requirements of NFPA 70.

605.17.1 Refrigerant detection system. The machinery room shall be provided with a refrigerant detection system. The refrigerant detection system shall be in accordance with Section 605.8 and all of the following:

1. The detectors shall activate at or below a refrigerant concentration of 25 percent of the LFL.
2. Upon activation, the detection system shall activate the emergency ventilation system in Section 605.17.2
3. The detection, signaling and control circuits shall be supervised.

605.13, 605.16, 605.17 continues

605.13, 605.16, 605.17 continued

605.17.2 Emergency ventilation system. An emergency ventilation system shall be provided at the minimum exhaust rate specified in ASHRAE 15 or Table 605.17.2 Shut down of the emergency ventilation system shall be by manual means.

TABLE 605.17.2 Minimum Exhaust Rate

REFRIGERANT	Q (m^3/sec)	Q (cfm)
R32	15.4	32,600
R143A	13.6	28,700
R444A	6.46	13,700
R444B	10.6	22,400
R445A	7.83	16,600
R446A	23.9	50,700
R447A	23.8	50,400
R451A	7.04	15,000
R451B	7.05	15,000
R1234yf	7.80	16,600
R1234ze(E)	5.92	12,600

605.17.3 Emergency ventilation system discharge. The point of discharge to the atmosphere shall be located outside of the structure at not less than 15 feet (4572 mm) above the adjoining grade level and not less than 20 feet (6096 mm) from any window, ventilation opening or exit.

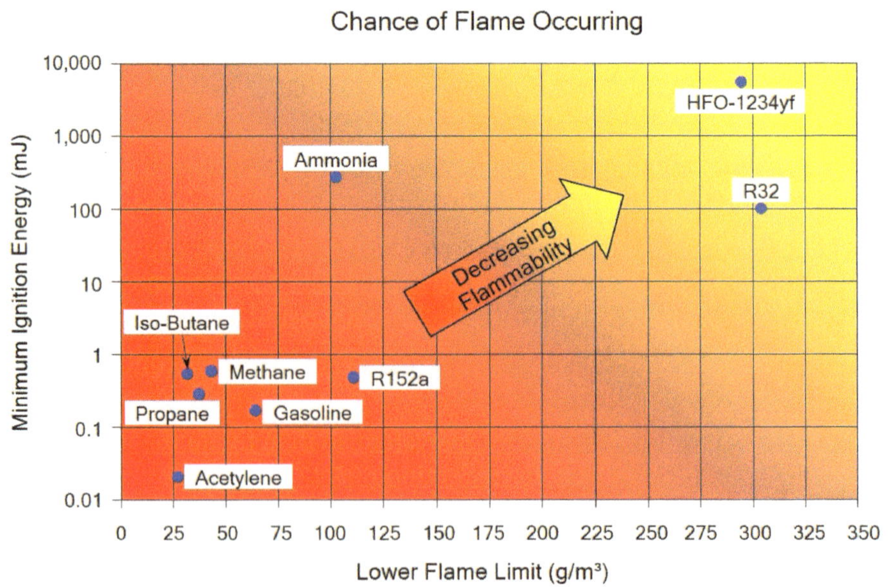

This graph compares the minimum energy required for ignition to occur. HFO-1234yf is difficult to ignite and classified as a Group A2L refrigerant.

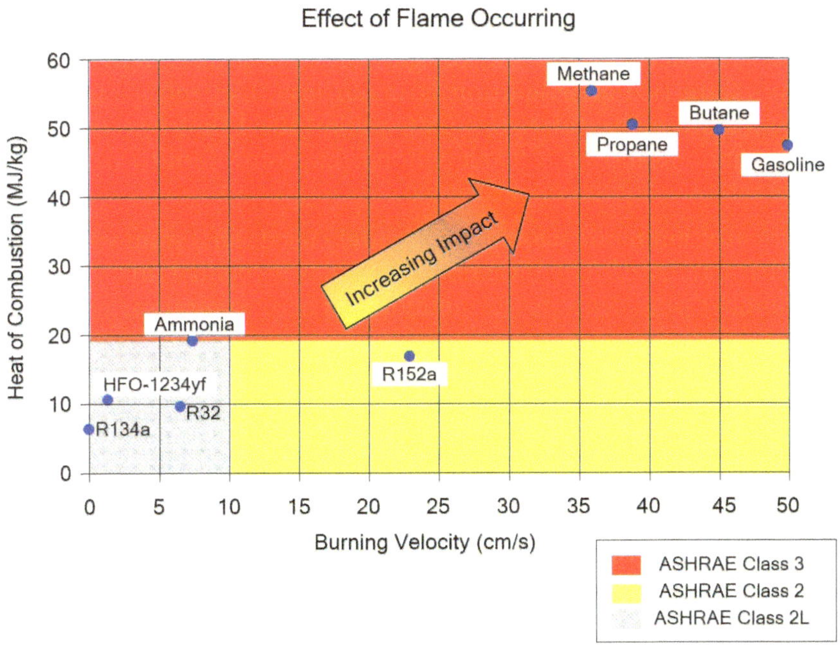

This graph compares the burning velocity after ignition occurs. HFO-1234yf is classified as a Group A2L refrigerant. It takes over 5,000 times more energy to ignite HFO-1234yf as compared to R152a, a Group 2 refrigerant.

CHANGE SIGNIFICANCE: In order to address documented changes in the earth's atmosphere, Federal environmental protection agencies have begun limiting the use of refrigerants that have high global warming-potential properties. The refrigerant manufacturing industry has responded by developing a number of replacement refrigerants. One of the characteristics of most of these products is that they are minimally flammable as compared to other flammable refrigerants. (See figures above.) Different protection schemes exist for lesser risk products which are not as restrictive as those for other more flammable refrigerants.

The change addresses a new class of refrigerants with very low flammability characteristics called Group A2L refrigerants. The change does not affect existing requirements for flammable and/or toxic refrigerant products. This change simply allows Group A2L refrigerants in refrigeration machinery rooms to comply with the current code requirements for Group 1, Division 2 electrical systems, or install detection and ventilation systems to mitigate the hazard.

The IFC, IMC, and ASHRAE have historically classified refrigerant flammability as Group 1 (nonflammable), Group 2 (moderately flammable), and Group 3 (highly flammable). The newly developed and environmentally preferred refrigerants present a lower flammability hazard than traditional Class 2 refrigerants; therefore, a new classification, Group 2L, was established.

In addition to flammability, refrigerants are classified as either nontoxic (Group A) or toxic (Group B). The new refrigerants are primarily classified as Group A2L—nontoxic, mildly flammable. The 2L refrigerants

605.13, 605.16, 605.17 continues

605.13, 605.16, 605.17 continued

have a burning velocity of less than 10 cm/sec when tested in accordance with ASHRAE Standard 34, "Designation and Safety Classification of Refrigerants." The energy required for ignition is very high, and the subsequent pressure rise after ignition is much less than refrigerants with higher flammability characteristics. IMC Table 1103.1 (Refrigeration Classification, Amount and OEL [Occupational Exposure Limit]) recognizes 2L refrigerants as a sub-class of Group 2.

Historically, ammonia was the only widely used refrigerant with these burning characteristics. The IFC and other codes recognize this, and have made a number of exceptions for ammonia. These exceptions are based on the fact that its burning characteristics reduce the risk of ignition, and the risk of damage if ignition does occur is greatly reduced. These risks are further reduced by code requirements for early detection and ventilation to dilute the ammonia concentration in air.

Detection and ventilation approaches are similar to many other hazardous operations, including repair garages, hazardous materials storage and use, compressed gases, corrosives, and other hazardous materials. This is a common and accepted mitigation scheme within the IFC and other codes.

The ventilation rates included in Table 606.17.2 are based on research that shows those rates will maintain a safe environment when more than 90 percent of the refrigerant is released by a leak.

608.3

Nonmetallic Cooking Oil Storage Tanks

CHANGE TYPE: Modification

CHANGE SUMMARY: Adds listing and capacity requirements for cooking oil storage.

2018 CODE: **608.3 Nonmetallic storage tanks.** Nonmetallic cooking oil storage tanks <u>shall be listed in accordance with UL 2152 and</u> shall be installed in accordance with the tank manufacturer's instructions. ~~and shall also comply with all of the following:~~

~~1. Tanks shall be listed for use with cooking oil, including maximum temperature to which the tank will be exposed during use.~~

~~2.~~ Tank capacity shall not exceed 200 gallons (757 L) per tank.

CHANGE SIGNIFICANCE: Fresh and waste cooking oils are normally classified as a Class IIIB combustible liquid. Storage tanks for cooking operations are regulated in Section 608.2, for metallic tanks, and Section 608.3, for nonmetallic tanks.

Metallic tanks over 60 gallons are required to be listed, and the quantity is not limited. Nonmetallic tanks over 60 gallons are required to be listed to UL 2152 and are limited to a maximum of 200 gallons.

UL 2152, "Outline of Investigation for Special Purpose Nonmetallic Containers & Tanks for Specific Combustible or Noncombustible Liquids" was developed to evaluate the construction and performance of nonmetallic tanks for the storage of new and waste cooking oil, among other applications. The fire code already required nonmetallic tanks to be listed for use with cooking oil, this change identifies the standard used to list these systems.

The exceptions in Section 5001.1 and the footnotes to Table 5003.1.1(1) do not exempt cooking oil storage. Table 5003.1.1(1) limits the quantity of Class IIIB combustible liquids to 13,200 gallons, unless the building is sprinklered, then the quantity is unlimited. Cooking oil would be included in determining this limit. There is no limit on the number of tanks, but a typical installation will consist of at least two tanks, one for fresh oil and one for waste oil.

Photo courtesy of Restaurant Technologies, Inc.

Typically, two tanks are installed for cooking oil storage; one for fresh oil, and one for waste oil.

807.1, 807.2, 807.5

Combustible Decorative Materials

International Code Council®

While the tree and glass globes are noncombustible decorative materials, the fabric draped across the ceiling is a combustible decoration and is limited to 10 percent of the ceiling area.

CHANGE TYPE: Modification

CHANGE SUMMARY: The limitations on decorative combustible materials are clarified as to where they apply.

2018 CODE: ~~**807.1 General.** Combustible decorative materials, other than decorative vegetation, shall comply with Sections 807.2 through 807.5.6.~~

~~**807.2**~~ **807.1** ~~**Limitations**~~ **General.** The following requirements shall apply to all occupancies:

1. Furnishings or decorative materials of an explosive or highly flammable character shall not be used.
2. Fire-retardant coatings in existing buildings shall be maintained so as to retain the effectiveness of the treatment under service conditions encountered in actual use.
3. Furnishings or other objects shall not be placed to obstruct exits, access thereto, egress therefrom or visibility thereof.
4. The permissible amount of decorative vegetation and noncombustible decorative materials shall not be limited.

~~**807.3**~~ **807.2 Combustible decorative materials.** In ~~other than Group I-3~~ Groups A, B, E, I, M and R-1 and in dormitories in Group R-2, curtains, draperies, fabric hangings and other similar combustible decorative materials suspended from walls or ceilings shall comply with Section ~~807.4~~ 807.3 and shall not exceed 10 percent of the specific wall or ceiling area to which such materials are attached.

Fixed or movable walls and partitions, paneling, wall pads and crash pads applied structurally or for decoration, acoustical correction, surface insulation or other purposes shall be considered to be *interior finish*, shall comply with Section 803 and shall not be considered *decorative materials* or furnishings.

Exceptions:

1. In auditoriums in Group A, the permissible amount of curtains, draperies, fabric hangings and similar combustible decorative material suspended from walls or ceilings shall not exceed 75 percent of the aggregate wall area where the building is equipped throughout with an *approved automatic sprinkler system* in accordance with Section 903.3.1.1, and where the material is installed in accordance with Section ~~803.13~~ 803.15 of the *International Building Code*.
2. In Group R-2 dormitories, within sleeping units and dwelling units, the permissible amount of curtains, draperies, fabric hangings and other similar decorative materials suspended from walls or ceilings shall not exceed 50 percent of the aggregate wall areas where the building is equipped throughout with an *approved automatic sprinkler system* installed in accordance with Section 903.3.1.
3. In Group B and M occupancies, the amount of combustible fabric partitions suspended from the ceiling and not supported by

the floor shall comply with Section ~~807.4~~ 807.3 and shall not be limited.

4. The 10-percent limit shall not apply to curtains, draperies, fabric hangings and similar combustible decorative materials used as window coverings.

807.5 Occupancy-based requirements. ~~In occupancies specified, combustible decorative materials not complying with Section 807.3~~ Occupancies shall comply with Sections 807.5.1 through 807.5.6.

CHANGE SIGNIFICANCE: The requirements for combustible decorative materials have been revised to add clarity in their application. Section 807.1 is revised by deleting the previous requirement which excluded decorative vegetation from the limitations. The list of items is not changed, but now Section 807.1 applies to all decorative materials and to all occupancies.

New Section 807.2 is revised to reinstate a list of specific occupancy classifications that previously appeared in the 2012 code. It is difficult to justify the restrictions in these requirements in factories and storage occupancies. For example, in a Group S-1 occupancy these combustible decorations could represent a much lower hazard than the products stored in the facility, and the facility is designed to handle this higher hazard. Application to Groups A, B, E, I, M, R-1, and dormitories in Group R-2 places these limitations into facilities with a higher life hazard and into the facilities where these hazards exist.

Exception 4 has been added to Section 807.2 to remove the limitations on window coverings. Virtually every hotel room in the world has a window. The *International Building Code* requires natural light in hotel rooms and many hotel rooms use glazing for the entire exterior wall. Yet, nearly every hotel provides black out curtains as a convenience to the customer. Under the 2015 code this practice was not allowed because even though the window coverings met the flame spread requirements, they were still limited to 10 percent of the wall area. The exclusion of window coverings allows this practice to continue. Note that this exception does not exclude the requirement for flame spread testing in Section 807.3. It only exempts the window coverings from the 10-percent limitation.

Section 807.5 has been revised because the allowance to exempt occupancies listed in this section from their applicable special regulations based on complying with Section 807.2 is potentially dangerous. For example, based on the strict reading of Section 807.5 in the 2015 IFC, restrictions on the use of foam plastics in Group A are currently waived if you comply with Section 807.3. This was never the intent, and this revision corrects this section. As written now, the occupancy based requirements in Section 807.5.1 through 807.5.6 apply in all situations, and are in addition to the requirements in Section 807.2.

807.4

Artificial Decorative Vegetation

International Code Council®

This artificial vegetation is not required to comply with the flammability ratings since the wreath is less than 50 percent of the door and the flower arrangement has dimensions less than 3 feet.

CHANGE TYPE: Modification

CHANGE SUMMARY: The limitations on decorative combustible materials are clarified as to where they apply.

2018 CODE:

SECTION 807
DECORATIVE MATERIALS ~~OTHER THAN~~ AND ARTIFICIAL DECORATIVE VEGETATION IN NEW AND EXISTING BUILDINGS

807.4 Artificial decorative vegetation. Artificial decorative vegetation shall comply with this section and with the requirements of Sections 806.2 and 806.3. Natural decorative vegetation shall comply with Section 806.

Exception: Testing of artificial vegetation is not required in Group I-1; Group I-2, Condition 1; Group R-2; Group R-3; or Group R-4 occupancies equipped throughout with an *approved automatic sprinkler system* installed in accordance with Section 903.3.1, where such artificial vegetation complies with the following:

1. Wreaths and other decorative items on doors shall not obstruct the door operation and shall not exceed 50 percent of the surface area of the door.
2. Decorative artificial vegetation shall be limited to not more than 30 percent of the wall area to which it is attached.
3. Decorative artificial vegetation not on doors or walls shall not exceed 3 feet (914 mm) in any dimension.

~~806.2~~ 807.4.1 ~~Artificial vegetation~~ Flammability. Artificial decorative vegetation shall meet the flame propagation performance criteria of Test Method 1 or Test Method 2, as appropriate, of NFPA 701. Meeting ~~the flame propagation performance~~ such criteria ~~of Test Method 1 or Test Method 2, as appropriate, of NFPA 701~~ shall be documented and certified by the manufacturer in an *approved* manner. Alternatively, the artificial decorative vegetation shall be tested in accordance with NFPA 289, using the 20 kW ignition source, and shall have a maximum heat release rate of 100 kW.

807.4.2 Electrical fixtures and wiring on artificial vegetation. The use of unlisted electrical wiring and lighting on artificial decorative vegetation shall be prohibited. The use of electrical wiring and lighting on artificial trees constructed entirely of metal shall be prohibited.

CHANGE SIGNIFICANCE: These revisions all address artificial decorative vegetation. The first change separates natural vegetation from artificial vegetation. Section 806 now deals with natural vegetation and Section 807 deals with artificial vegetation along with other decorative materials.

Owners and operators of Group I-1, assisted living facilities; Group I-2, Condition 1, nursing homes; Group R-2, congregate living facilities;

Group R-3, boarding houses and congregate care facilities; and Group R-4, group homes, often seek to create a home-like environment. Residents desire to personalize their space by displaying decorative items. There is often a desire for display of seasonal decorative items, like a wreath, and it is not always possible to determine compliance with NFPA 701 or NFPA 289. However, large artificial vegetation items that do not comply with either of the fire tests have been shown to generate massive amounts of heat when they burn. Therefore, the exception to Section 807.4, Item 1 allows these small decorative items in sprinklered facilities without the need for fire testing. Note that the reference for sprinkler design is to Section 903.3.1. This means that a fire sprinkler system can be designed to NFPA 13, "Standard for the Installation of Sprinkler Systems;" NFPA 13R, "Standard for the Installation of Sprinkler Systems in Low-Rise Residential Occupancies;" or NFPA 13D, "Standard for the Installation of Sprinkler Systems in One- and Two-Family Dwellings and Manufactured Homes."

These revisions provide guidance to what has previously been allowed under the allowance of "limited quantities." There are three limiting factors for artificial vegetation.

1. Wreaths on entry doors cannot cover more than 50 percent of the door and cannot interfere with the door operation.
2. Garlands or other items that might be attached to a wall are limited to a maximum of 30 percent of the wall area.
3. Other artificial decorative vegetation is limited to a maximum dimension of three feet. Larger items, such as an artificial tree or larger artificial plant would need to comply with the NFPA testing requirements.

Section 807.4.1 is moved from Section 806.2 since it deals specifically with artificial vegetation. This section is clarified with no technical change.

Section 807.4.2 is added to deal with electrical wiring and lighting on artificial vegetation. This section states that the wiring must be listed. If the artificial vegetation is metallic, then wiring and electrical equipment, whether listed or not, is prohibited.

Section 807.4 states that the artificial vegetation must still comply with Sections 806.2 and 806.3. These two sections deal with proximity to the egress path and ignition sources. These requirements apply whether the vegetation is natural or artificial.

901.4.6.1, 901.4.6.2, 901.4.6.3, 901.4.6.4

Fire Pump and Fire Sprinkler Riser Rooms

International Code Council®

Fire sprinkler riser room is identified and a lock box is provided to allow fire department access.

CHANGE TYPE: Addition

CHANGE SUMMARY: Additional requirements have been added for automatic sprinkler system riser rooms and fire pump rooms.

2018 CODE: **901.4.6.1 Access.** Automatic sprinkler system risers, fire pumps and controllers shall be provided with ready access. Where located in a fire pump room or automatic sprinkler system riser room, the door shall be permitted to be locked provided the key is available at all times.

901.4.6.2 Marking on access doors. Access doors for automatic sprinkler system riser rooms and fire pump rooms shall be labeled with an approved sign. The lettering shall be in a contrasting color to the background. Letters shall have a minimum height of 2 inches (51 mm) with a minimum stroke of $^{3}/_{8}$ inch (10 mm).

901.4.6.3 Environment. Automatic sprinkler system riser rooms and fire pump rooms shall be maintained at a temperature of not less than 40ºF (4ºC). Heating units shall be permanently installed.

901.4.6.4 Lighting. Permanently installed artificial illumination shall be provided in the automatic sprinkler system riser rooms and fire pump rooms.

CHANGE SIGNIFICANCE: This code change adds additional criteria to fire pump rooms and fire sprinkler riser rooms. The code does not require such rooms to be provided, but where they are provided, the rooms must meet the requirements in Section 901.4.6 of the IFC and Section 902.1 of the *International Building Code.*

Fire pump and fire sprinkler riser rooms must be readily accessible at all times, but can be locked if a key is available. This could be accomplished using a key box as provided in Section 506.1.

The temperature in the room must be maintained at 40º Fahrenheit or higher to ensure the piping and components do not freeze. This same temperature threshold is required in NFPA 20, "Standard for the Installation of Stationary Pumps for Fire Protection." NFPA 13, "Standard for the Installation of Sprinkler Systems" also specifies the same temperature threshold when determining the need for a dry-pipe sprinkler system.

Permanent lighting is required in fire pump rooms and fire sprinkler riser rooms.

CHANGE TYPE: Addition

901.6.2

Integrated Fire Protection System Testing

CHANGE SUMMARY: Test criteria has been added to the code with a reference to NFPA 4 to ensure that where multiple fire protection systems or life safety systems are integrated, that the acceptance process and subsequent testing must evaluate all of the integrated systems as a whole.

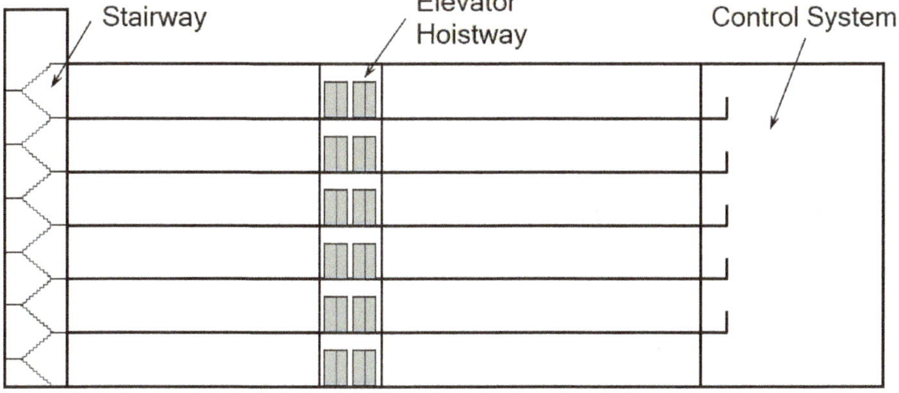

Smoke detectors in the elevator lobbies are connected to the elevator controls. Smoke detection in the atrium activates the smoke control system and closes the fire doors separating the atrium from the remainder of the building. The ventilation system for pressurized stairways and pressurized hoistways is activated by smoke detection, fire alarm, or waterflow on the sprinkler system. These various systems must be integrated to function together properly.

2018 CODE: **901.6.2 Integrated testing.** Where two or more fire protection or life safety systems are interconnected, the intended response of subordinate fire protection and life safety systems shall be verified when required testing of the initiating system is conducted. In addition, integrated testing shall be performed in accordance with Sections 901.6.2.1 and 901.6.2.2.

901.6.2.1 High-rise buildings. For high-rise buildings, integrated testing shall comply with NFPA 4, with an integrated test performed prior to issuance of the certificate of occupancy and at intervals not exceeding 10 years, unless otherwise specified by an integrated system test plan prepared in accordance with NFPA 4. If an equipment failure is detected during integrated testing, a repeat of the integrated test shall not be required, except as necessary to verify operation of fire protection or life safety functions that are initiated by equipment that was repaired or replaced.

901.6.2.2 Smoke control systems. Where a fire alarm system is integrated with a smoke control system as outlined in Section 909, integrated testing shall comply with NFPA 4, with an integrated test performed prior to issuance of the certificate of occupancy and at intervals not exceeding 10 years, unless otherwise specified by an integrated system test plan prepared in accordance with NFPA 4. If an equipment failure is detected during integrated testing, a repeat of the integrated test shall not be required, except as necessary to verify operation of fire protection or life safety functions that are initiated by equipment that was repaired or replaced.

901.6.2 continues

901.6.2 continued

SECTION 202 DEFINITIONS

INTEGRATED TESTING (FIRE PROTECTION AND LIFE SAFETY SYSTEM). A testing procedure to establish the operational status, interaction and coordination of two or more fire protection and safety systems.

SUBORDINATE (FIRE PROTECTION AND LIFE SAFETY SYSTEM). A system that is activated by another fire protection or life safety system. For example, where a fire alarm system activates a smoke removal or elevator recall system, the smoke removal or elevator recall system is considered to be "subordinate" to the fire alarm system.

CHANGE SIGNIFICANCE: The code clearly intends to require proper operation of all fire protection and life safety systems within a building. In many cases, such as monitoring the fire sprinkler system with the fire alarm system and notifying a monitoring service when a fire sprinkler operates, the integration is relatively simple. However in some cases, such as a fire alarm system initiating a complex combination of door closures, damper operations, elevator captures, and fan activations in a high-rise building, the integration can be highly complex and in most cases it involves the coordination of many different trades, control units and systems. This code revision is created to ensure that required testing of integrated features is scaled in a manner that is reasonable for a wide range of applications.

NFPA 4, "Standard for Integrated Fire Protection and Life Safety System Testing" is a new standard that deals with the testing of integrated systems. NFPA 4 is now referenced in Sections 901.6.2.1 and 901.6.2.2. These two sections require compliance with NFPA 4 for the more complex systems found in high-rise buildings and smoke control systems. Section 901.6.2 specifies that in all buildings where there are multiple systems, they must be integrated, but it does not require compliance with NFPA 4 for these less complex designs.

This creates a two-tiered approach to integrated testing which is designed to address both the highly complex scenarios and those that are not so complex where it is a simpler task to verify the functionality of integrated fire protection and life safety systems.

It should also be noted that the requirement established by Section 901.6.2 intends to convey that required testing of integrated features should follow a top down approach, where testing of an initiating device or control warrants verification of subordinate systems or functions, but not necessarily the reverse. For example, where a smoke detector is intended to trigger an automatic damper (the damper being a "subordinate" device to the smoke detector), the test protocol should verify that the damper system responds as intended when the smoke detector activates. However, if a damper control system is tested simply to exercise dampers and to verify that dampers are operating, such testing should not require integrated testing of the initiating smoke detector or fire alarm system.

Lastly, with respect to high-rise buildings and smoke control systems, the text makes it clear that in addition to the initial acceptance testing prior to Certificate of Occupancy, a repeat of the entire integrated testing process is required at 10 year intervals.

901.8.2

Removal of Occupant-use Hose Lines

CHANGE TYPE: Modification

CHANGE SUMMARY: Authorizes code official to allow the removal of occupant-use hose lines.

2018 CODE: **901.8.2 Removal of existing occupant-use hose lines.** The *fire code official* is authorized to permit the removal of existing occupant-use hose lines where ~~all~~ both of the following conditions exist:

~~1. Installation is not required by this code or the *International Building Code*.~~

1. The hose line would not be utilized by trained personnel or the fire department.
2. The remaining outlets are compatible with local fire department fittings.

CHANGE SIGNIFICANCE: This change provides additional guidance to the fire code official for those circumstances where occupant-use hose lines may be removed. The occupant use hose is typically $1^1/_2$-inch hose with a nozzle which is part of a Class II standpipe system. Class II hose connections are not required when the building is sprinklered; most new construction is now required to be sprinklered. The *International Fire Code* and *International Building Code* only require the Class II standpipe system for stages with greater than 1,000 square feet.

The benefit of fire sprinklers in the building outweighs the benefit of occupant-use hose lines. Based on the discussions at the code hearings, it was determined that in some occupancies it is considered a better strategy to have the occupants evacuate and let the fire sprinklers operate, rather than have the occupants attempt extinguishment. The occupant-use hose lines should remain in place where there is an on-site fire brigade at the facility. These personnel are trained to combat the fire, while the other occupants are evacuating.

International Code Council®

This hose cabinet previously contained hose and nozzle for occupant use. These items have been removed and the remaining connection is compatible with the fire department hose fittings.

903.2.1

Sprinklers in Group A Occupancies

CHANGE TYPE: Clarification

CHANGE SUMMARY: Clarifies the requirements for fire sprinkler protection in Group A occupancies.

2018 CODE: **903.2.1 Group A.** An *automatic sprinkler system* shall be provided throughout buildings and portions thereof used as Group A occupancies as provided in this section. ~~For Group A-1, A-2, A- 3 and A-4 occupancies, the automatic sprinkler system shall be provided throughout the story where the fire area containing the Group A-1, A-2, A-3 or A-4 occupancy is located, and throughout all stories from the Group A occupancy to, and including, the levels of exit discharge serving the Group A occupancy. For Group A-5 occupancies, the automatic sprinkler system shall be provided in the spaces indicated in Section 903.2.1.5.~~

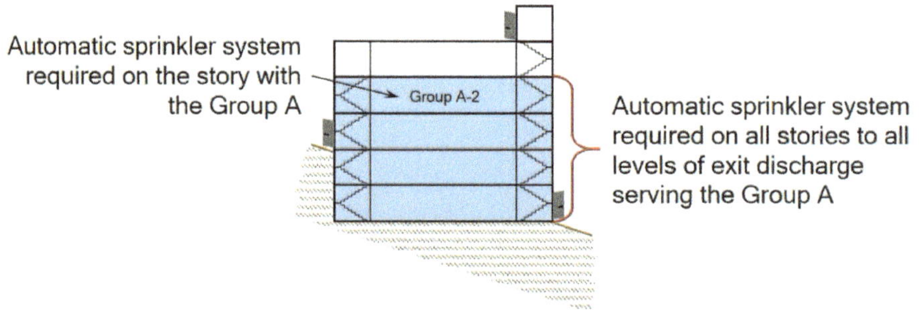

Where a Group A occupancy is located on a level other than the level of exit discharge, an automatic sprinkler system is required in the entire story and all stories to and including all levels of exit discharge serving the Group A occupancy.

903.2.1.1 Group A-1. An *automatic sprinkler system* shall be provided ~~for fire areas~~ throughout stories containing Group A-1 occupancies and ~~intervening floors~~ throughout all stories from the Group A-1 occupancy to and including the levels of ~~the building~~ exit discharge serving that occupancy where one of the following conditions exists:

1. The *fire area* exceeds 12,000 square feet (1115 m²).
2. The *fire area* has an *occupant load* of 300 or more.
3. The *fire area* is located on a floor other than a *level of exit discharge* serving such occupancies.
4. The *fire area* contains a multiple-theater complex.

903.2.1.2 Group A-2. An *automatic sprinkler system* shall be provided ~~for fire areas~~ throughout stories containing Group A-2 occupancies and ~~intervening floors~~ throughout all stories from the Group A-2 occupancy to and including the levels of ~~the building~~ exit discharge serving that occupancy where one of the following conditions exists:

1. The *fire area* exceeds 5,000 square feet (464 m²).
2. The *fire area* has an *occupant load* of 100 or more.
3. The *fire area* is located on a floor other than a *level of exit discharge* serving such occupancies.

903.2.1.3 Group A-3. An *automatic sprinkler system* shall be provided ~~for fire areas~~ throughout stories containing Group A-3 occupancies and ~~intervening floors~~ throughout all stories from the Group A-3 occupancy to and including the levels of ~~the building~~ exit discharge serving that occupancy where one of the following conditions exists:

1. The *fire areaa* exceeds 12,000 square feet (1115 m²).
2. The *fire area* has an *occupant load* of 300 or more.
3. The *fire area* is located on a floor other than a *level of exit discharge* serving such occupancies.

903.2.1.4 Group A-4. An *automatic sprinkler system* shall be provided ~~for fire areas~~ throughout stories containing Group A-4 occupancies and ~~intervening floors~~ throughout all stories from the Group A-4 occupancy to and including the levels of ~~the building~~ exit discharge serving that occupancy where one of the following conditions exists:

1. The *fire area* exceeds 12,000 square feet (1115 m²).
2. The *fire area* has an *occupant load* of 300 or more.
3. The *fire area* is located on a floor other than a *level of exit discharge* serving such occupancies.

903.2.1.5 Group A-5. An *automatic sprinkler system* shall be provided for all enclosed Group A-5 ~~occupancies in the following areas: concession stands, retail areas, press boxes and other~~ accessory use areas in excess of 1,000 square feet (93 m²).

903.2.1.5.1 Spaces under grandstands or bleachers. Enclosed spaces under *grandstands* or *bleachers* shall be equipped with an *automatic sprinkler system* in accordance with Section 903.3.1.1 where either of the following exist:

1. The enclosed area is 1,000 square feet (93 m²) or less and is not constructed in accordance with Section 1029.1.1.1.
2. The enclosed area exceeds 1,000 square feet (93 m²).

CHANGE SIGNIFICANCE: These changes clarify the language where fire sprinkler systems are required in Group A occupancies and other spaces within the same building. Furthermore, it details requirements for the fire sprinkler system to protect spaces on the level of exit discharge.

Previous code language created an inconsistency among Sections 903.2.1.1 through 903.2.1.4. Section 903.2.1 stated sprinklers were required on the story with the Group A occupancy and on all floors to, and including, the level of exit discharge serving the Group A occupancy. Sections 903.2.1.1 through 903.2.1.4 used the term "intervening" floors when referring to the same requirement. However, the definition of "intervene"

903.2.1 continues

903.2.1 continued

is "to occur or be between two things." Stories "intervening" or "between" the Group A occupancy and the level of exit discharge did not include sprinklers on the level of exit discharge. This proposal corrects that inconsistency and relocates that important code provision into each section where it applies.

Section 903.2.1.5.1 was added to correlate the sprinkler requirements with the requirements for enclosed accessory spaces under grandstands and bleachers requirements. There are no technical changes from what existed in the 2015 editions of the codes. All spaces under bleachers or grandstands, except toilet rooms and ticket booths less than 100 square feet, must be separated by 1-hour fire barriers and horizontal assemblies. Section 903.2.1.5.1 Item 1 provides an alternative to use fire sprinkler protection in lieu of fire-resistance-rated construction if the enclosed area is less than 1,000 square feet.

Furthermore, IFC Section 304.1 was revised to read:

304.1.3.1 Space underneath ~~seats~~ grandstands and bleachers. ~~Spaces underneath grandstand and bleacher seats shall be kept free from combustible and flammable materials. Except where enclosed in not less than 1- hour fire-resistance-rated construction in accordance with the International Building Code, spaces underneath grandstand and bleacher seats shall not be occupied or utilized for purposes other than means of egress.~~ Spaces underneath grandstands and bleachers shall not be occupied or utilized for purposes other than means of egress except where equipped with an automatic sprinkler system in accordance with Section 903.2.1.5.1, or separated with fire barriers and horizontal assemblies in accordance with Section 1029.1.1.1.

Concurrently, Section 1029.1.1.1 was revised to read:

1029.1.1.1 Spaces under grandstands and bleachers. ~~Where spaces~~ Spaces under *grandstands* or *bleachers* ~~are used for purposes other than ticket booths less than 100 square feet (9.29 m^2) and toilet rooms, such spaces~~ shall be separated by *fire barriers* complying with Section 707 of the *International Building Code* and *horizontal assemblies* complying with Section 711 of the *International Building Code* with not less than 1-hour fire-resistance-rated construction.

Exceptions:

1. Ticket booths less than 100 square feet (9 m^2) in area.
2. Toilet rooms.
3. Other accessory use areas 1,000 square feet (93 m^2) or less in area and equipped with an automatic sprinkler system in accordance with Section 903.3.1.1.

Section 903.2.1.5.1, Item 2 requires enclosed spaces greater than 1,000 square feet to be equipped with automatic sprinklers. For these spaces over 1,000 square feet, the combined requirements of Sections 903.2.1.5.1 and 1029.1.1.1 results in 1-hour construction and an automatic sprinkler system.

903.2.3

Sprinklers in Group E Occupancies

CHANGE TYPE: Modification

CHANGE SUMMARY: Provides occupant load threshold for automatic sprinkler system requirements in Group E occupancies.

2018 CODE: **903.2.3 Group E.** An *automatic sprinkler system* shall be provided for Group E occupancies as follows:

1. Throughout all Group E *fire areas* greater than 12,000 square feet (1115 m^2) in area.
2. The Group E fire area is located on a floor other than a level of exit discharge serving such occupancies.

    Exception: In buildings where every classroom has not fewer than one exterior exit door at ground level, an automatic sprinkler system is not required in any area below the lowest level of exit discharge serving that area.

~~2. Throughout every portion of educational buildings below the lowest level of exit discharge serving that portion of the building.~~

Exception: ~~An automatic sprinkler system is not required in any area below the lowest level of exit discharge serving that area where every classroom throughout the building has not fewer than one exterior exit door at ground level.~~

3. The Group E fire area has an occupant load of 300 or more.

903.2.3 continues

International Code Council®

Group E occupancies now require an automatic sprinkler system when the occupant load in a fire area is 300 or more.

903.2.3 continued

CHANGE SIGNIFICANCE: This change requires fire sprinklers in all Group E fire areas having an occupant load of 300 persons or more, regardless of fire area size. Using Table 1004.5 "Maximum Floor Area Allowances per Occupant," a Group E fire area of 12,000 square feet could have an occupant load of 600 persons (12,000 square feet ÷ 20 square feet per person = 600). This occupant load is twice the occupant load threshold where Group A occupancies would require automatic sprinklers.

Many schools serve multiple purposes, including parent/teacher conference night, open houses for new or prospective parents and students, political elections, social and club meetings, religious gatherings and adult education classes. If these uses are to be allowed, then the actual use of the school becomes more similar to a Group A occupancy, not Group E.

The occupant load threshold of 300 is a common value used throughout the code. It is the threshold for sprinklers in Group A-1, A-3, and A-4 occupancies; for manual fire alarm systems in Group A occupancies (Section 907.2.1); assembly main exit provisions (Section 1029.2); horizontal building separation allowances (IBC Section 510.2); Class C interior finishes in nonsprinklered Group A-1 or A-2 rooms; or enclosed spaces (Table 803.3 Footnote e) and door locks and latches in Group A (Section 1010.1.9.4).

Item 2 clarifies that fire sprinklers are required when the Group E occupancy is located on a floor other than the level of exit discharge. The exception is technically unchanged from the 2015 edition of the code, but has been revised for editorial clarity.

903.3.1.1.2

Sprinklers in Bathrooms in Group R Occupancies

CHANGE TYPE: Modification

CHANGE SUMMARY: Removes fire sprinkler requirements from small bathrooms in Group R-4 occupancies.

2018 CODE: **903.3.1.1.2 Bathrooms.** In Group R occupancies, ~~other than Group R-4 occupancies~~, sprinklers shall not be required in bathrooms that do not exceed 55 square feet (5 m^2) in area and are located within individual *dwelling units* or *sleeping units*, provided that walls and ceilings, including the walls and ceilings behind a shower enclosure or tub, are of noncombustible or limited-combustible materials with a 15-minute thermal barrier rating.

CHANGE SIGNIFICANCE: This change allows the omission of fire sprinklers from small bathrooms in Group R-4 occupancies. These occupancies provide dwelling space for five to 16 persons on a 24-hour basis and provide some sort of custodial care. These occupancies include alcohol and drug centers, assisted living facilities, congregate care facilities, group homes, halfway houses, residential board and care facilities and social rehabilitation facilities.

Sprinklers were omitted from small bathrooms in these occupancies to be consistent among all Group R occupancies.

Table 722.6.2(1) in the 2018 IBC can be used to determine the fire-resistance rating of walls constructed with membranes installed on framing members that are spaced not more than 16 inches on-center. Based on the table, a 15-minute thermal barrier can be achieved with a wood frame wall with a single layer of $^1/_2$-inch gypsum.

Photo courtesy of Assistant Fire Marshal Fulton Cochran, Clark County Building and Fire Prevention Services

Small bathrooms in Group R-4 occupancies do not need to be protected with fire sprinklers.

903.3.1.2.1

Sprinklers Beneath Balconies

CHANGE TYPE: Modification

CHANGE SUMMARY: Correlates automatic sprinkler system requirements in Chapter 9 with IBC Chapter 7 for exterior balconies of Group R occupancies.

2018 CODE: **903.3.1.2.1 Balconies and decks.** Sprinkler protection shall be provided for exterior balconies, decks and ground floor patios of *dwelling units* and *sleeping units* where either of the following conditions exists:

1. The building is of Type V construction, provided that there is a roof or deck above.
2. Exterior balconies, decks and ground floor patios of dwelling units and sleeping units are constructed in accordance with Section 705.2.3.1, Exception 3 of the *International Building Code.*

Sidewall sprinklers that are used to protect such areas shall be permitted to be located such that their deflectors are within 1 inch (25 mm) to 6 inches (152 mm) below the structural members and a maximum distance of 14 inches (356 mm) below the deck of the exterior balconies and decks that are constructed of open wood joist construction.

CHANGE SIGNIFICANCE: The change correlates the IFC and IBC regarding fire sprinkler protection for exterior balconies, decks and ground floor patios outside dwelling and sleeping units.

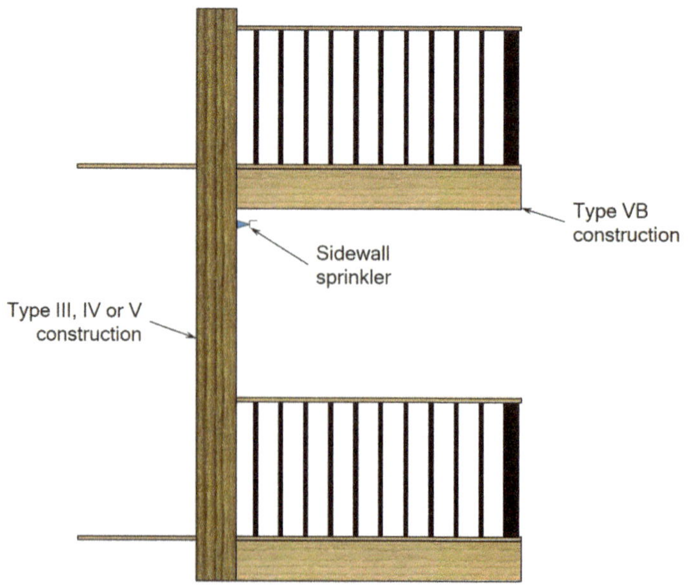

In buildings of Type III, IV or V construction, the exterior balcony can be Type V construction with no fire-resistance-rating if sprinklers are installed to protect the balcony.

The *International Building Code*, Section 705.2.3.1, Exception 3 states "Balconies and similar projections on buildings of Type III, IV and V construction shall be permitted to be of Type V construction, and shall not be required to have a fire-resistance rating where sprinkler protection is extended to these areas." The new Item 2 recognizes this requirement in the IBC and specifies that sprinkler protection is required over the balcony when this construction provision is utilized.

Fire sprinklers for exterior balconies, decks and ground floor patios with a roof above are intended to prevent exterior or exposure fires from spreading vertically or horizontally along the outside wall. Numerous significant losses have occurred in Group R multifamily occupancies where fires have spread up combustible exterior siding and entered unprotected openings, roof vents, or roof structures.

903.3.1.2.3

Protection of Attics in Group R Occupancies

CHANGE TYPE: Addition

CHANGE SUMMARY: Sprinkler protection of acceptable alternative methods for the protection of attics are now addressed for mid-rise buildings housing multifamily occupancies and equipped with an NFPA 13R sprinkler system.

2018 CODE: **903.3.1.2.3 Attics.** Attic protection shall be provided as follows:

1. Attics that are used or intended for living purposes or storage shall be protected by an *automatic sprinkler system.*
2. Where fuel-fired equipment is installed in an unsprinklered attic, not fewer than one quick-response intermediate temperature sprinkler shall be installed above the equipment.
3. Where located in a building of Type III, Type IV or Type V construction designed in accordance with Section 510.2 or 510.4 of the *International Building Code*, attics not required by Item 1 to have sprinklers shall comply with one of the following if the roof assembly is located more than 55 feet (16 764 mm) above the lowest level of required fire department vehicle access:
    3.1. Provide *automatic sprinkler system* protection.
    3.2. Construct the attic using noncombustible materials.
    3.3. Construct the attic using fire-retardant-treated wood complying with Section 2303.2 of the *International Building Code.*
    3.4. Fill the attic with noncombustible insulation.

The height of the roof assembly shall be determined by measuring the distance from the lowest required fire vehicle access road surface adjacent to the building to the eave of the highest

International Code Council®

The upper portion of this building is Group R-2 and can be protected by an NFPA 13R sprinkler system, but when the building exceeds 55 feet in height the attic must be protected by one of four options.

pitched roof, the intersection of the highest roof to the exterior wall or the top of the highest parapet, whichever yields the greatest distance. For the purpose of this measurement, required fire vehicle access roads shall include only those roads that are necessary for compliance with Section 503.

4. Group R-4, Condition 2 occupancy attics not required by Item 1 to have sprinklers shall comply with one of the following:
   4.1. Provide *automatic sprinkler system* protection.
   4.2. Provide a heat detection system throughout the attic that is arranged to activate the building fire alarm system.
   4.3. Construct the attic using noncombustible materials.
   4.4. Construct the attic using fire-retardant-treated wood complying with Section 2303.2 of the *International Building Code.*
   4.5. Fill the attic with noncombustible insulation.

CHANGE SIGNIFICANCE: Recently, major fires in combustible, multiple-story multifamily dwellings resulted in questions about the effectiveness of automatic fire sprinklers in these buildings. These buildings often were protected by fire sprinkler systems designed and installed in accordance with NFPA 13R, "Standard for the Installation of Sprinkler Systems in Low-rise Residential Buildings" to protect occupant lives, but were not intended for property protection. Many of the fires started on the building exterior, spread vertically into unprotected combustible attics and then literally burned from the top down.

This change addresses concerns about fire sprinkler control in taller residential buildings and "pedestal" buildings consisting of a noncombustible lower level topped by combustible living units.

Attic protection is required when the roof assembly is located more than 55 feet above the lowest level of required fire department vehicle access. This means pedestal buildings that exceed a height of 55 feet above grade plane, including the pedestal, will be affected by the change. Some buildings with fewer stories that are located on sloped lots with fire department vehicle access roads required along the lower-elevation portion of the perimeter also may be affected. Note that the measurement is from the lowest level of a required fire apparatus access road, not from grade plane. The access road can often be at an elevation below grade plane.

The intent of identifying "required" fire apparatus access is to make it clear that, simply because access is available on an adjacent road or parking lot, the road need not be considered in the height measurement unless it is required as part of satisfying the code requirement for vehicle access to the building.

The permissible attic protection options generally are modeled after existing Section 903.2.8.3 that was adopted in the 2015 edition for Group R-4, Condition 2 buildings.

903.3.1.2.3 continues

903.3.1.2.3 continued

Allowances for noncombustible construction materials, fire-retardant-treated wood, and filling with noncombustible insulation have for many years been permitted by NFPA 13, "Standard for the Installation of Sprinkler Systems" as an acceptable alternative to installing sprinklers in concealed spaces in otherwise fully-sprinklered buildings.

The change relocates the existing requirements in Section 903.2.8.3 for enhanced attic protection in Group R-4, Condition 2 occupancies to the new Section 903.3.1.2.3 so all supplemental and alternative protection requirements for NFPA 13R sprinkler systems are consolidated in one location.

The existing requirements for Group R-4, Condition 2 also were revised to clarify that, in an otherwise unsprinklered attic, the entire attic would not require sprinklers based on the presence of fuel-fired equipment, provided that the roof was not greater than 55 feet above fire apparatus access roads. Instead, Section 6.6.6.1 of the 2016 NFPA 13R requires that a sprinkler be installed only over the equipment.

903.3.3

Sprinkler Obstructions

CHANGE TYPE: Modification

CHANGE SUMMARY: The code now directs the user to the sprinkler design standard to address sprinkler obstructions.

2018 CODE: 903.3.3 Obstructed locations. Automatic sprinklers shall be installed with ~~due~~ regard to obstructions that will delay activation or obstruct the water distribution pattern <u>and shall be in accordance with the applicable *automatic sprinkler system* standard that is being used.</u> Automatic sprinklers shall be installed in or under covered kiosks, displays, booths, concession stands or equipment that exceeds 4 feet (1219 mm) in width. Not less than a 3-foot (914 mm) clearance shall be maintained between automatic sprinklers and the top of piles of *combustible fibers*.

> **Exception:** Kitchen equipment under exhaust hoods protected with a fire-extinguishing system in accordance with Section 904.

CHANGE SIGNIFICANCE: The change provides guidance to refer to the applicable sprinkler standard when dealing with obstructions to sprinkler water discharge. The previous language "installed with due regard to obstructions" was vague and provided no guidance.

This revision sends the code user to the specific design standard regarding sprinkler obstructions. For example, Chapter 8 of NFPA 13, "Standard for Installation of Sprinkler Systems" addresses obstructions to water spray. It specifically addresses obstructions greater than 4 feet wide, open floor grating, distances from beams and soffits, floor mounted obstructions, ceiling mounted obstructions, and many other situations. NFPA 13 Section 8.12 contains additional obstruction provisions specific to early suppression fast-response sprinklers (ESFR) sprinklers.

The intent of Section 903.3.3 has not changed. It is to have the sprinklers installed so that obstructions are minimized and the water spray can adequately protect the hazard.

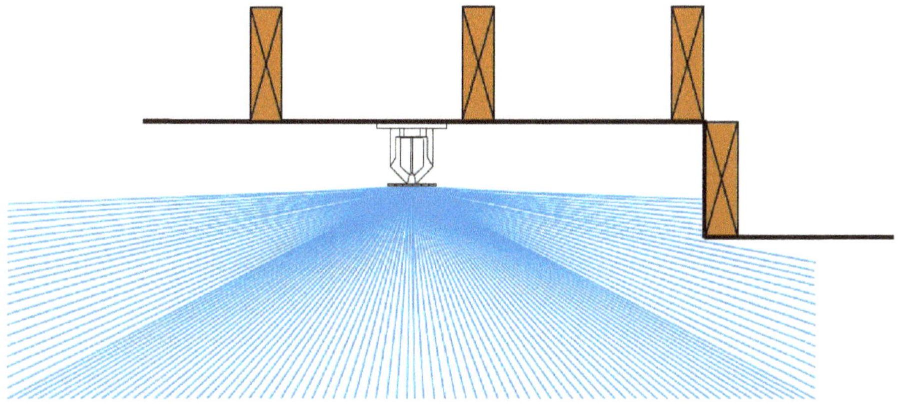

Fire sprinklers must be installed to either avoid obstructions, or adjusted so that the obstructions do not negatively impact the spray pattern.

904.12

Commercial Cooking Operations

CHANGE TYPE: Modification

CHANGE SUMMARY: The installation of fire-extinguishing systems as protection for commercial cooking operations must now comply with NFPA 96. In addition, commercial cooking systems are now permitted to be protected with a water mist fire-extinguishing system.

2018 CODE: **904.12 Commercial cooking systems.** The automatic fire-extinguishing system for commercial cooking systems shall be of a type recognized for protection of commercial cooking equipment and exhaust systems of the type and arrangement protected. Preengineered automatic dry- and wet-chemical extinguishing systems shall be tested in accordance with UL 300 and *listed* and *labeled* for the intended application. Other types of automatic fire-extinguishing systems shall be *listed* and *labeled* for specific use as protection for commercial cooking operations. The system shall be installed in accordance with this code, NFPA 96, its listing and the manufacturer's installation instructions. Automatic fire-extinguishing systems of the following types shall be installed in accordance with the referenced standard indicated, as follows:

1. Carbon dioxide extinguishing systems, NFPA 12.
2. *Automatic sprinkler systems*, NFPA 13.
3. Automatic water mist systems, NFPA 750.

~~3.~~4. Foam-water sprinkler system or foam-water spray systems, NFPA 16.

~~4.~~5. Dry-chemical extinguishing systems, NFPA 17.

~~5.~~6. Wet-chemical extinguishing systems, NFPA 17A.

Exception: Factory-built commercial cooking recirculating systems that are tested in accordance with UL 710B and *listed*, *labeled* and installed in accordance with Section 304.1 of the *International Mechanical Code.*

Photo courtesy of CaptiveAire

These cooking appliances are protected with an automatic water mist system.

CHANGE SIGNIFICANCE: A new reference is added to NFPA 96, "Standard for Ventilation Control and Fire Protection of Commercial Cooking Operations" for commercial cooking operations. Having a direct reference from the *International Fire Code* to NFPA 96, similar to the references to other NFPA standards for specific system types provides for a more efficient reference to applicable standards instead of sending the designer or code official through the *International Mechanical Code.* NFPA 96 is added to IFC Chapter 80 as a referenced standard. The purpose of NFPA 96 is to reduce the potential fire hazard of cooking operations and is relevant to the safety and operation of commercial cooking systems.

Water mist systems are added to the list of fire-extinguishing systems applicable for protection of commercial cooking systems. Several systems have been tested and approved for protection of industrial cooking systems. NFPA 96 includes water mist extinguishing systems as an acceptable fire-extinguishing system for commercial cooking.

Water mist systems consist of devices that emit a specific spray, or mist, that absorbs heat, displaces oxygen, and blocks heat transfer to control, suppress, or extinguish fires.

FM Global has approved water mist systems from multiple companies for use in industrial oil cookers. The industrial oil cookers are commonly used in the food industry, containing large amounts of oil, up to 5000 gallons, and operating at high temperatures. The FM Global approvals also apply to "downscaled" applications similar to what might be found in a commercial restaurant.

904.13

Domestic Cooking in Institutional Occupancies

Photo courtesy of Denlar Fire Protection

This domestic cooking appliance is protected with a UL 300A fire-extinguishing system.

CHANGE TYPE: Modification

CHANGE SUMMARY: Requires an automatic fire-extinguishing system to protect domestic cooking appliances in care facilities.

2018 CODE: **904.13 Domestic cooking systems.** ~~**in Group I-2 Condition 1.** In Group I-2 Condition 1 occupancies where cooking facilities are installed in accordance with Section 407.2.6 of the *International Building Code*, the domestic cooking hood provided over the cooktop or range shall be equipped with an automatic fire-extinguishing system of a type recognized for protection of domestic cooking equipment. Preengineered automatic extinguishing systems shall be tested in accordance with UL 300A and listed and labeled for the intended application. The system shall be installed in accordance with this code, its listing and the manufacturer's instructions.~~ Cooktops and ranges installed in the following occupancies shall be protected in accordance with Section 904.13.1:

1. In Group I-1 occupancies where domestic cooking facilities are installed in accordance with Section 420.8 of the *International Building Code.*
2. In Group I-2, Condition 1 occupancies where domestic cooking facilities are installed in accordance with Section 407.2.6 of the *International Building Code.*
3. In Group R-2 college dormitories where domestic cooking facilities are installed in accordance with Section 420.10 of the *International Building Code.*

~~**904.13.1 Manual system operation and interconnection.** Manual actuation and system interconnection for the hood suppression system shall be in accordance with Sections 904.12.1 and 904.12.2, respectively.~~

904.13.1 Protection from fire. Cooktops and ranges shall be protected in accordance with Section 904.13.1.1 or 904.13.1.2.

904.13.1.1 Automatic fire-extinguishing system. The domestic recirculating or exterior vented cooking hood provided over the cooktop or range shall be equipped with an approved automatic fire-extinguishing system complying with the following:

1. The automatic fire-extinguishing system shall be of a type recognized for protection of domestic cooking equipment. Preengineered automatic fire-extinguishing systems shall be listed and labeled in accordance with UL 300A and installed in accordance with the manufacturer's instructions.
2. Manual actuation of the fire-extinguishing system shall be provided in accordance with Section 904.12.1.
3. Interconnection of the fuel and electric power supply shall be in accordance with Section 904.12.2.

904.13.1.2 Ignition prevention. Cooktops and ranges shall include burners that have been tested and listed to prevent ignition of cooking oil with burners turned on to their maximum heat settings and allowed to operate for 30 minutes.

~~**904.13.2 Portable fire extinguishers for domestic cooking equipment in Group I-2 Condition 1.** A portable fire extinguisher complying with Section 906 shall be installed within a 30-foot (9144 mm) distance of travel from domestic cooking appliances.~~

CHANGE SIGNIFICANCE: This change correlates fire protection requirements for domestic cooking in alcohol and drug treatment centers, assisted living facilities, group homes, halfway houses, social rehabilitation facilities and similar occupancies (Group I-1) for more than 16 persons who live there on a 24-hour basis in a supervised environment and receive custodial care.

Occupants of these facilities are often capable of preparing small meals in cooking spaces less sophisticated than a full-scale commercial cooking operation. Even though a smaller scale, hazards still exist and those not involved with the cooking operation still need to be protected. Section 904.13.1 provides an option, either provide a fire-extinguishing system or install a cooking appliance meeting specific heating limitations.

Section 904.13.1.1 provides the criteria for design of an automatic fire-extinguishing system. The system must be designed and tested to UL 300A "Outline of Investigation for Extinguishing System Units for Residential Range Top Cooking Surfaces." This standard is not equivalent to the UL 300 standard for commercial cooking facilities.

Option 2 is in Section 904.13.2. This section requires the cooktops and ranges to meet specific heating limitations. These cooking appliances have listed ignition-resistant burners that do not allow cooking oils to ignite during testing.

Recent work by the Fire Protection Research Foundation confirms that burners meeting these specifications are highly unlikely to ignite cooking materials. Their report can be found at: http://www.nfpa.org/research/fire-protection-research-foundation/reports-andproceedings/other-research-topics/analytical-modeling-of-pan-and-oil-heating-on-an-electric-coil-cooktop. UL 858, "Standard for Safety for Household Electric Ranges" was recently revised to include a new Section 60A that evaluates the ability of burners to not ignite cooking oil.

Section 904.13.2 is shown as deleted, but it is only deleted from this section. The requirement for portable fire extinguishers near domestic cooking appliances is now found in Section 906.1, Item 2.

904.14, Table 901.6.1

Aerosol Fire-extinguishing Systems

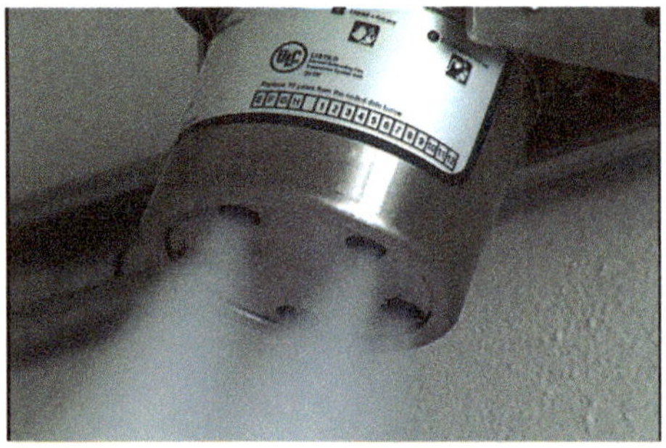

Photo courtesy of Fireaway Inc.

This is an aerosol fire-extinguishing device that is electrically operated.

CHANGE TYPE: Addition

CHANGE SUMMARY: Requires automatic fire suppression in domestic cooking systems in care facilities.

2018 CODE: **904.14 Aerosol fire-extinguishing systems.** Aerosol fire extinguishing systems shall be installed, periodically inspected, tested and maintained in accordance with Sections 901 and 904.4, NFPA 2010, and in accordance with their listing.

Such devices and appurtenances shall be listed and installed in conformance with manufacturer's instructions.

904.14.1 Maintenance. Not less than semiannually, an inspection shall be conducted by a trained person to assess whether the system is in working order. Not less than annually, a certified fire suppression contractor trained and having knowledge of the installation, operation and maintenance of the specific fire-extinguishing system shall inspect, test, service and maintain such system in accordance with this section and the manufacturer's specifications and servicing manuals.

TABLE 901.6.1 Fire Protection System Maintenance Standards

SYSTEM	STANDARD
Portable fire extinguishers	NFPA 10
Carbon dioxide fire-extinguishing system	NFPA 12
Halon 1301 fire-extinguishing systems	NFPA 12A
Dry-chemical extinguishing systems	NFPA 17
Wet-chemical extinguishing systems	NFPA 17A
Water-based fire protection systems	NFPA 25
Fire alarm systems	NFPA 72
Smoke and heat vents	NFPA 204
Water-mist systems	NFPA 750
Clean-agent extinguishing systems	NFPA 2001
Aerosol fire-extinguishing systems	NFPA 2010

CHANGE SIGNIFICANCE: NFPA 2010, "Standard for Fixed Aerosol Fire-extinguishing Systems" was first published in 2006. Since then, the International Code Council Evaluation Service® (ICC-ES®) published the ICC-ES Acceptance Criteria for Fixed Condensed Aerosol Fire-Extinguishing Systems, AC432.

In 2014, the ICC-ES published evaluation report ESR-3230 for an aerosol fire-extinguishing system in compliance with IFC 2009 and IFC 2012 as an alternative to Section 904.9 for halon fire-extinguishing systems.

Aerosol fire-extinguishing systems are now recognized by the code as an alternative fire-extinguishing system and NFPA 2010, "Standard for Fixed Aerosol Fire-Extinguishing Systems" has been added to Chapter 80.

Condensed aerosol fire suppression systems used as total flooding systems for the protection of Class A (surface), Class B, and Class C hazards can reduce construction, installation, and maintenance costs compared with existing fire-extinguishing systems. This technology does not use compressed gas cylinders nor pressure rated piping. Generally, these systems are electrically operated when integrated with an approved fire alarm and releasing control systems and releasing panels, or are deployed as automatic stand-alone fire-extinguishing units.

As there are no piping distribution systems required, no special storage requirements for compressed gas bottles, and the ability of the flooding agent to protect areas with limited leakage, the construction costs involving these systems are typically lower than conventional chemical and gas fire-extinguishing systems requiring gas pressure.

The technology remains effective even with leakage in the space, and therefore offers an alternative technology for installations involving reconstruction, or new construction in areas with special hazards, or where total room integrity construction is problematic.

As the units are designed to be disposable, generally they require very little maintenance and consequently offer low, long-term life cycle costs to facility operators.

Every six months a visual inspection is required to determine if the equipment appears to be functional and ready. Annually, a trained technician will inspect, service and maintain the system as required in NFPA 2010 and the manufacturer's instructions.

905.3.1

Class III Standpipes

CHANGE TYPE: Modification

CHANGE SUMMARY: Standpipe systems are now required in buildings four or more stories in height. In addition, a Class I standpipe is allowed in Group B and Group E occupancies rather than a Class III.

2018 CODE: **905.3.1 Height.** Class III standpipe systems shall be installed throughout buildings where any of the following conditions exist:

1. Four or more stories are above or below grade plane.
2. The floor level of the highest story is located more than 30 feet (9144 mm) above the lowest level of the fire department vehicle access.~~, or where the~~
3. The floor level of the lowest story is located more than 30 feet (9144 mm) below the highest level of fire department vehicle access.

Exceptions:

1. Class I standpipes are allowed in buildings equipped throughout with an *automatic sprinkler system* in accordance with Section 903.3.1.1 or 903.3.1.2.
2. Class I standpipes are allowed in Group B and E occupancies.

~~2.~~3. Class I manual standpipes are allowed in open parking garages where the highest floor is located not more than 150 feet (45 720 mm) above the lowest level of fire department vehicle access.

~~3.~~4. Class I manual dry standpipes are allowed in open parking garages that are subject to freezing temperatures, provided that the hose connections are located as required for Class II standpipes in accordance with Section 905.5.

~~4.~~5. Class I standpipes are allowed in *basements* equipped throughout with an *automatic sprinkler system*.

6. Class I standpipes are allowed in buildings where occupant-use hose lines will not be utilized by trained personnel or the fire department.

~~5.~~7. In determining the lowest level of fire department vehicle access, it shall not be required to consider either of the following:

   ~~5.1.~~7.1. Recessed loading docks for four vehicles or less.

   ~~5.2.~~7.2. Conditions where topography makes access from the fire department vehicle to the building impractical or impossible.

CHANGE SIGNIFICANCE: The three changes in Section 905.3.1 clarify the Class III standpipe requirements based on building height, allow modifications to the class of standpipe system in Group B and Group E occupancies, and allow the elimination of occupant-use hose lines in specified situations.

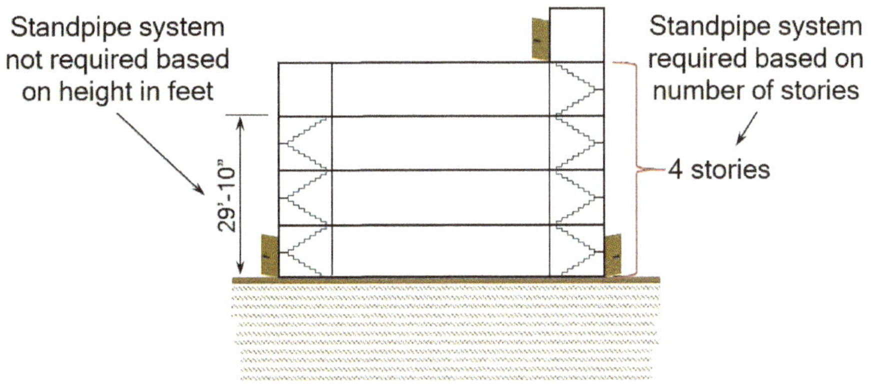

Even though the highest floor level is less than 30 feet above fire department vehicle access, the building requires a standpipe system because it is four stories or more.

Previous requirements for fire department use standpipes were based solely on the height in feet between the lowest level of fire department access and the floor level of the highest or lowest story. The change in Section 905.3.1 retains the height criteria in feet, but also adds the criteria that any building four or more stories above or below grade is required to have a standpipe system.

For example, a standpipe is required for any four-story building with story heights of 10 feet on flat ground where fire department vehicle access is at the lowest level, essentially the first or ground floor.

In other conditions, where the elevation of the fire apparatus access road does not align with the first floor level, the lowest stairway landing also likely does not align with the road. However, the interior stairway and landing arrangements where standpipes normally are located don't substantially change based on whether the access road level is a few feet above or below the stairs. The height requirement of 30 feet is intended to address those circumstances.

Exception 2 is added to address the issue of occupant-use hose lines in Group B and E occupancies. It removes the occupant-use hose lines from these occupancies by allowing a Class I standpipe.

Exception 6 is added which allows the installation of a Class I standpipe in lieu of a Class III standpipe where the facility will not have trained personnel to utilize the occupant-use hose lines and the fire department will not utilize the hose lines. Based on Exception 1 and the new Exception 2, most buildings will install a Class I standpipe rather than a Class III standpipe. The requirement for a Class III standpipe occurs at four stories or 30 feet above or below the level of fire department vehicle access. All new buildings will also be sprinklered when they reach these same thresholds, except Groups B, F-2, S-2, and U occupancies. Exception 2 now eliminates Class II, or occupant-use hose lines, from Group B. The only occupancies left are Groups F-2, S-2, and U.

905.3.1 continues

905.3.1 continued

In recent years, many fire safety and evacuation plans have all but abandoned the use of occupant-use hose lines in their training to the building occupants and employees. The primary focus of the training is evacuation. Fire behavior has changed dramatically in the past several decades because of changes in fire loading. This has created fires that develop faster, create more heat in most situations, and produce greater amounts of toxic smoke. As a result, the ability for occupants to safety and effectively utilize occupant-use hose lines without the protection of firefighting gear and respiratory protection has been greatly minimized. Where personnel will not be trained, the code now allows the elimination of the occupant-use hose lines.

Note that the exceptions, including Exception 6, only apply where the standpipe is required in Section 905.3.1. Other sections in the code also require standpipes, but are not affected by these exceptions. For example, Section 905.3.4 requires a Class III standpipe on stages greater than 1,000 square feet.

905.4

Class I Standpipe Hose Connections

CHANGE TYPE: Modification

CHANGE SUMMARY: Allows a modification of hose connection locations for Class I standpipes serving open stairways.

2018 CODE: 905.4 Location of Class I standpipe hose connections. Class I standpipe hose connections shall be provided in all of the following locations:

1. In every required *interior exit stairway*, a hose connection shall be provided for each story above and below grade plane. Hose connections shall be located at ~~an intermediate~~ the main floor landing ~~between stories,~~ unless otherwise *approved* by the *fire code official*.

    Exception: A single hose connection shall be permitted to be installed in the open corridor or open breezeway between open stairs that are not greater than 75 feet (22 860 mm) apart.

2. *No change in text...*
3. *No change in text...*
4. *No change in text...*
5. *No change in text...*
6. *No change in text...*

905.4 continues

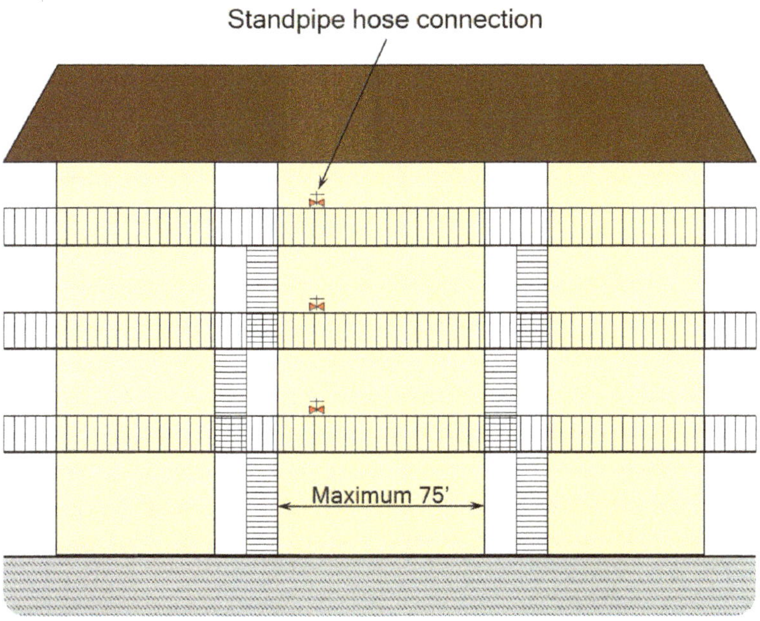

A single hose connection can be installed between open stairways connected by open breezeways provided the stairways are no further than 75 feet apart.

905.4 continued

CHANGE SIGNIFICANCE: The change in Item 1 brings the hose valve placement requirements into correlation with NFPA 14, "Standard for the Installation of Standpipe and Hose Systems" that requires hose valves to be located at the main floor landing of each story, typically at the same elevation as the exit doors into the stairs.

While the installation of hose valves at intermediate landings may seem more convenient for fire fighter operational staging, that arrangement typically requires separate risers to be run for sprinklers and standpipes. This increases the cost and requires significantly more materials to achieve code compliance. There is no evidence to support the value of retaining the hose valves at the intermediate landing.

The new exception to Item 1 recognizes that there is no significant value to having two standpipes located at opposite ends of an open breezeway or corridor that connects to open stairs since both standpipes are essentially sharing the same environmental space. With stairways located not more than 75 feet apart, and a hose line typically consisting of 150 feet, the two hose connections would overlap each other by a distance of 75 feet. This redundancy is not needed and therefore not required. The requirement is to provide a hose connection somewhere between the two stairways. It does not need be in the middle, it could be located at one stairway, or closer to one stairway than the other.

905.11

Locking Caps on Standpipe Outlets

CHANGE TYPE: Modification

CHANGE SUMMARY: This revision authorizes the code official to require locking caps on dry standpipe hose connection outlets.

2018 CODE: **905.11 Locking standpipe outlet caps**. The *fire code official* is authorized to require locking caps on the outlets on dry standpipes where the responding fire department carries key wrenches for the removal that are compatible with locking FDC connection caps.

CHANGE SIGNIFICANCE: This change will allow the code official to require locking caps on dry standpipe hose station outlets. These outlets are vulnerable to vandalism when trash and other debris are stuffed into the waterway.

Note that this is applicable to dry standpipe connections, not wet standpipes. In addition to the brass caps being stolen and sold for scrap, dry standpipe valves can also be opened and debris lodged inside. When debris or other obstructions are put into the hose station outlet, they may block water from flowing into fire fighter hose lines when needed.

Dry standpipes also are susceptible to having one or more hose station valves left open. When this occurs, adequate flow and pressure cannot be maintained at the hose station outlet(s) in use. The locking caps increases security and dependability for the dry standpipe system.

906.1

Portable Fire Extinguishers

International Code Council®

Where each classroom in a Group E occupancy is provided with a fire extinguisher, additional fire extinguishers in the corridors are not required.

CHANGE TYPE: Modification

CHANGE SUMMARY: Provides schools options for fire extinguisher placement.

2018 CODE: **906.1 Where required.** Portable fire extinguishers shall be installed in all of the following locations:

1. In new and existing Group A, B, E, F, H, I, M, R-1, R-2, R-4 and S occupancies.

   Exceptions:

   1. In Group R-2 occupancies, portable fire extinguishers shall be required only in locations specified in Items 2 through 6 where each dwelling unit is provided with a portable fire extinguisher having a minimum rating of 1-A:10-B:C.
   2. In Group E occupancies, portable fire extinguishers shall be required only in locations specified in Items 2 through 6 where each classroom is provided with a portable fire extinguisher having a minimum rating of 2-A:20-B:C.

2. Within 30 feet (9144 mm) distance of travel from commercial cooking equipment and from domestic cooking equipment in Group I-1; I-2, Condition 1; and R-2 college dormitory occupancies.
3. In areas where flammable or combustible liquids are stored, used or dispensed.
4. On each floor of structures under construction, except Group R-3 occupancies, in accordance with Section 3315.1.
5. Where required by the sections indicated in Table 906.1.
6. Special-hazard areas, including but not limited to laboratories, computer rooms and generator rooms, where required by the *fire code official*.

CHANGE SIGNIFICANCE: This change provides an option for schools to provide portable fire extinguishers in classrooms in lieu of centrally located or distributed extinguishers throughout the school buildings. The fire extinguishers in the common areas are more susceptible to mischief and are not available when the school is in lockdown. Locating a fire extinguisher within each classroom improves security of this equipment and also places the extinguishers within reach while a lockdown is in progress. While the potential is low for a lockdown and fire to occur simultaneously, this solution is an option that provides for that event.

Schools are now required to develop lock down plans to protect students and faculty from intruders. The plans effectively prevent access to portable extinguishers normally located in hallways during lockdown situations.

907.1.2

Fire Alarm Construction Documents

CHANGE TYPE: Modification

CHANGE SUMMARY: Aligns requirements for fire alarm plans and documentation requirements with NFPA 72.

2018 CODE: **907.1.2 Fire alarm shop drawings.** Shop drawings for fire alarm systems shall be prepared in accordance with NFPA 72 and submitted for review and approval prior to system installation.~~, and shall include, but not be limited to, all of the following where applicable to the system being installed:~~

~~1. A floor plan that indicates the use of all rooms.~~

~~2. Locations of alarm-initiating devices.~~

~~3. Locations of alarm notification appliances, including candela ratings for visible alarm notification appliances.~~

~~4. Design minimum audibility level for occupant notification.~~

~~5. Location of fire alarm control unit, transponders and notification power supplies.~~

~~6. Annunciators.~~

~~7. Power connection.~~

~~8. Battery calculations.~~

~~9. Conductor type and sizes.~~

~~10. Voltage drop calculations.~~

~~11. Manufacturers' data sheets indicating model numbers and listing information for equipment, devices and materials.~~

~~12. Details of ceiling height and construction.~~

~~13. The interface of fire safety control functions.~~

~~14. Classification of the supervising station.~~

CHANGE SIGNIFICANCE: This change removes the IFC's prescriptive list of items that must be included in a set of fire alarm plan review submittals and refers the designer to NFPA 72, *National Fire Alarm Code* for guidance. This reference ensures there are no conflicts between the code language and the design standard.

NFPA 72, Section 7.4, requires at least 31 items to be included in the construction documentation. One of these items is a written description of the system design and sequence of operation. Even though the 2015 code states "...including, but not limited to..." it adds clarity to simply refer to NFPA 72 rather than provide a list which obviously does not include all the documentation that is needed. All of items previously listed in the code are included in the required documentation in NFPA 72.

International Code Council®

The fire alarm control unit controls the fire alarm system and its location must be shown on the construction documents.

907.2.1

Fire Alarms in Group A Occupancies

CHANGE TYPE: Modification

CHANGE SUMMARY: A new fire alarm threshold has been added for Group A occupancies where an occupant load of 100 or more is located on a level other than the level of exit discharge.

2018 CODE: **907.2.1 Group A.** A manual fire alarm system that activates the occupant notification system in accordance with Section 907.5 shall be installed in Group A occupancies where the occupant load due to the assembly occupancy is 300 or more, or where the Group A occupant load is more than 100 persons above or below the *lowest level of exit discharge*. Group A occupancies not separated from one another in accordance with Section 707.3.10 of the *International Building Code* shall be considered as a single occupancy for the purposes of applying this section. Portions of Group E occupancies occupied for assembly purposes shall be provided with a fire alarm system as required for the Group E occupancy.

> **Exception:** Manual fire alarm boxes are not required where the building is equipped throughout with an automatic sprinkler system installed in accordance with Section 903.3.1.1 and the occupant notification appliances will activate throughout the notification zones upon sprinkler water flow.

CHANGE SIGNIFICANCE: A manual fire alarm system is required in Group A occupancies where any of the following situations occur:

1. There is an aggregate occupant load of 300 or more.
2. There is an occupant load of 100 or more located above or below the level of exit discharge.

Group B and M occupancies are required to provide a fire alarm system when the occupant load exceeds 100 on floors other than the level of exit discharge. It was deemed appropriate to apply this same requirement in Group A occupancies.

A Group A occupancy on a floor other than the level of exit discharge is required to be equipped with an automatic sprinkler system, regardless of the occupant load. Now, when the Group A occupancy is on a floor

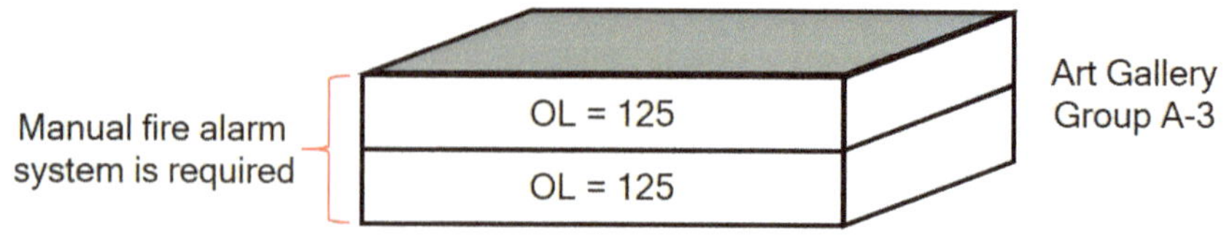

A fire alarm system is not required in Group A-3 with an aggregate occupant load of 250, but since the occupant load on the second floor exceeds 100, a fire alarm system is required.

other than the level of exit discharge and the occupant load exceeds 100 a manual fire alarm system will be added to the protection scheme. The exception states that the manual fire alarm boxes are not required when the building is sprinklered. IFC Section 907.2 states that when manual fire alarm boxes are allowed to be eliminated, at least one shall be provided in an approved location.

907.2.10 (deleted)

Group R-4 Fire Alarm System

International Code Council®

The requirement for a fire alarm system in Group R-4 occupancies has been deleted.

CHANGE TYPE: Deletion

CHANGE SUMMARY: Fire alarm systems are no longer required in Group R-4 occupancies.

2018 CODE: ~~**907.2.10 Group R-4.** Fire alarm systems and smoke alarms shall be installed in Group R-4 occupancies as required in Sections 907.2.10.1 through 907.2.10.3.~~

~~**907.2.10.1 Manual fire alarm system.** A manual fire alarm system that activates the occupant notification system in accordance with Section 907.5 shall be installed in Group R-4 occupancies.~~

Exceptions:

~~1. A manual fire alarm system is not required in buildings not more than two stories in height where all individual sleeping units and contiguous attic and crawl spaces to those units are separated from each other and public or common areas by not less than 1-hour fire partitions and each individual sleeping unit has an exit directly to a public way, egress court or yard.~~

~~2. Manual fire alarm boxes are not required throughout the building where all of the following conditions are met:~~

~~2.1. The building is equipped throughout with an automatic sprinkler system installed in accordance with Section 903.3.1.1 or 903.3.1.2.~~

~~2.2. The notification appliances will activate upon sprinkler water flow.~~

~~2.3. Not fewer than one manual fire alarm box is installed at an approved location.~~

~~3. Manual fire alarm boxes in resident or patient sleeping areas shall not be required at exits where located at all nurses' control stations or other constantly attended staff locations, provided such stations are visible and continuously accessible and that the distances of travel required in Section 907.4.2.1 are not exceeded.~~

~~**907.2.10.2 Automatic smoke detection system.** An automatic smoke detection system that activates the occupant notification system in accordance with Section 907.5 shall be installed in corridors, waiting areas open to corridors and habitable spaces other than sleeping units and kitchens.~~

Exceptions:

~~1. Smoke detection in habitable spaces is not required where the facility is equipped throughout with an automatic sprinkler system installed in accordance with Section 903.3.1.1.~~

~~2. An automatic smoke detection system is not required in buildings that do not have interior corridors serving sleeping units and where each sleeping unit has a means of egress door opening directly to an exit or to an exterior exit access that leads directly to an exit.~~

~~**907.2.10.3 Smoke alarms.** Single and multiple station smoke alarms shall be installed in accordance with Section 907.2.11.~~

CHANGE SIGNIFICANCE: The requirement to install a manual fire alarm system and an automatic smoke detection system in Group R-4 occupancies has been deleted. Group R-4 occupancies are group homes with residents capable of self-preservation. The environment is very similar to Group R-3 where the code does not require a fire alarm system.

The 2015 IFC requires a manual fire alarm system and smoke detection system in a Group R-4 facility with 16 or fewer residents. A Group R-4 is less of a hazard than a Group R-1 where the occupants may not be familiar with their surroundings, but yet the alarm systems are not required until the Group R-1 is three stories in height. A Group R-4 facility is where the residents are effectively working together similar to a single-family home rather than individual living units.

Single- and multiple-station smoke alarms are still required in Group R-4 occupancies as stated in Section 907.2.10.2.

907.5.2.2.4

Emergency Voice/ Alarm Communication System Captions

CHANGE TYPE: Modification

CHANGE SUMMARY: Large public venues are required to provide real-time captions that are integrated with the emergency voice/alarm communication system.

2018 CODE: **907.5.2.2.4 Emergency voice/alarm communication captions.** Where stadiums, arenas and grandstands ~~are required to caption~~ have 15,000 fixed seats or more and provide audible public announcements, ~~in accordance with Section 1108.2.7.3~~, the emergency voice/alarm communication system shall ~~be captioned~~ provide pre-recorded or real-time captions. Prerecorded or live emergency captions shall be from an *approved* location constantly attended by personnel trained to respond to an emergency.

CHANGE SIGNIFICANCE: This revision adds language from Section 1108.2.7.3 in the *International Building Code* to correlate the accessibility provisions with the requirements in Chapter 9. IBC Section 1108.2.7.3 requires real-time captions of all audible public announcements. This revision affords the code official and designers to design and enforce the requirements for emergency voice/alarm communication systems and coordinate with the requirements for accessibility.

This provides for a visual notification to the large audience in these venues.

International Code Council®

This venue, with over 15,000 fixed seats, requires captioned announcements from the emergency voice/alarm communications system.

910.5

Maintenance of Smoke and Heat Removal Equipment

CHANGE TYPE: Modification

CHANGE SUMMARY: Maintenance and testing frequencies for smoke and heat vents and mechanical smoke removal are specified in the code.

2018 CODE: **910.5 Maintenance and testing.** ~~Smoke~~ Maintenance and testing of smoke and heat vents and mechanical smoke removal systems shall be ~~maintained in an operative condition~~ in accordance with Sections 910.5.1 ~~or~~ and 910.5.2, ~~respectively.~~ A written record of inspection, testing and maintenance that includes the date, identification of personnel involved, any unsatisfactory result, corrective action taken and replaced parts shall be maintained on the premises.

910.5.1 Smoke and heat vents. Smoke and heat vents shall be maintained in an operative condition. Inspection, testing and maintenance shall be in accordance with NFPA 204 ~~and Section 910.5.1.1~~ except as follows:

1. Mechanically operated smoke and heat vents shall be inspected annually and operationally tested not less than every 5 years.
2. Gravity dropout smoke and heat vents shall be inspected annually.
3. Fused, damaged or painted fusible links shall be replaced.

~~**910.5.1.1 Fusible links.** Fusible links for smoke and heat vents shall be replaced whenever fused, damaged or painted.~~

910.5.2 Mechanical smoke removal systems. Mechanical smoke removal systems shall be maintained in accordance with NFPA 204 and the equipment manufacturer's ~~maintenance~~ instructions ~~and Sections 910.5.2.1 through 910.5.2.4~~ except as follows:

1. Systems shall be inspected and operationally tested annually.
2. Testing shall include the operation of all system components, controls and ancillary equipment, such as makeup air openings.
3. A written schedule for routine maintenance and operational testing shall be established and testing shall be conducted in accordance with the schedule.

~~**910.5.2.1 Frequency.** Systems shall be operationally tested not less than once per year. Testing shall include the operation of all system components, including control elements.~~

~~**910.5.2.2 Testing.** Operational testing of the mechanical smoke removal system shall include all equipment such as fans, controls and make-up air openings.~~

~~**910.5.2.3 Schedule.** A routine maintenance and operational testing program shall be initiated and a written schedule for routine maintenance and operational testing shall be established.~~

Photo courtesy of Bilco, Inc.

Each smoke and heat vent is to be inspected annually and tested at least every five years.

910.5 continues

910.5 continued

~~**910.5.2.4 Written record.** A written record of mechanical smoke exhaust system testing and maintenance shall be maintained on the premises. The written record shall include the date of the maintenance, identification of the servicing personnel and notification of any unsatisfactory condition and the corrective action taken, including parts replaced.~~

CHANGE SIGNIFICANCE: Smoke and heat vents, like other fire protection devices, need to be inspected, maintained and tested on a regular basis to ensure proper operation. NFPA 204, "Standard for Smoke and Heat Venting," is referenced in the 2015 IFC and contains requirements for inspection and testing. NFPA 204 requires annual inspection of smoke and heat vents and gravity drop out smoke and heat vents. Further, NFPA 204 requires annual testing of mechanically operated smoke and heat vents. The requirements for inspection and testing are provided in the IFC to provide the inspector with the needed information to require the inspection, testing and maintenance (ITM) of these devices. However, rather than testing all smoke and heat vents annually, this proposal revises that testing requirement to once every 5 years.

The second sentence in Section 910.5 addresses recordkeeping of inspections and maintenance. It requires that the records include not only the date and names of inspection/maintenance personnel, but also notes what repairs are made and what components are replaced.

Section 910.5.1 addresses ITM for smoke and heat vents, while Section 910.5.2 addresses mechanical smoke removal systems. Both types of smoke removal equipment must be inspected annually, but the testing frequencies are specifically different than those found in NFPA 204.

Since the inspection, testing and maintenance frequencies are specified in the code, the frequencies found in NFPA 204 are superseded by the code language. It was determined that a five-year schedule provides an adequate inspection frequency and annual testing of smoke and heat vents was not warranted.

Testing every smoke and heat vent annually is an aggressive schedule, and commonly not being enforced. By changing the frequency to five years, the equipment has a better chance of getting tested and enforced locally.

ITM for mechanical smoke removal systems remains at once per year.

CHANGE TYPE: Modification

CHANGE SUMMARY: Requirements for gas detection systems are clarified and consolidated in a new Section 916.

916

Gas Detection Systems

This gas detector is listed to detect natural gas or LP-gas.

2018 CODE:

SECTION 916 GAS DETECTION SYSTEMS

916.1 Gas detection systems. Gas detection systems required by this code shall comply with Sections 916.2 through 916.11.

916.2 Permits. Permits shall be required as set forth in Section 105.7.11.

916.2.1 Construction documents. Documentation of the gas detection system design and equipment to be used that demonstrates compliance with the requirements of this code shall be provided with the application for permit.

916.3 Equipment. Gas detection system equipment shall be designed for use with the gases being detected and shall be installed in accordance with manufacturer's instructions.

916.4 Power connections. Gas detection systems shall be permanently connected to the building electrical power supply or shall be permitted to be cord connected to an unswitched receptacle using an *approved* restraining means that secures the plug to the receptacle.

916.5 Emergency and standby power. Standby or emergency power shall be provided or the gas detection system shall initiate a trouble signal at an *approved* location if the power supply is interrupted.

916.6 Sensor locations. Sensors shall be installed in approved locations where leaking gases are expected to accumulate.

916.7 Gas sampling. Gas sampling shall be performed continuously. Sample analysis shall be processed immediately after sampling, except as follows:

1. For HPM gases, sample analysis shall be performed at intervals not exceeding 30 minutes.
2. For toxic gases that are not HPM, sample analysis shall be performed at intervals not exceeding 5 minutes, in accordance with Section 6004.2.2.7.
3. Where a less frequent or delayed sampling interval is *approved.*

916.8 System activation. A gas detection alarm shall be initiated where any sensor detects a concentration of gas exceeding the following thresholds:

1. For flammable gases, a gas concentration exceeding 25 percent of the lower flammable limit (LFL).
2. For nonflammable gases, a gas concentration exceeding one-half of the IDLH, unless a different threshold is specified by the section of this code requiring a gas detection system.

916 continues

916 continued

Upon activation of a gas detection alarm, alarm signals or other required responses shall be as specified by the section of this code requiring a gas detection system. Audible and visible alarm signals associated with a gas detection alarm shall be distinct from fire alarm and carbon monoxide alarm signals.

916.9 Signage. Signs shall be provided adjacent to gas detection system alarm signaling devices that advise occupants of the nature of the signals and actions to take in response to the signal.

916.10 Fire alarm system connections. Gas sensors and gas detection systems shall not be connected to fire alarm systems unless *approved* and connected in accordance with the fire alarm equipment manufacturer's instructions.

916.11 Inspection, testing and sensor calibration. Inspection and testing of gas detection systems shall be conducted not less than annually. Sensor calibration shall be confirmed at the time of sensor installation and calibration shall be performed at the frequency specified by the sensor manufacturer.

SECTION 202 DEFINITIONS

GAS DETECTION SYSTEM. ~~A gas detection system where the analytical instrument is maintained in continuous operation and sampling is performed without interruption. Analysis is allowed to be performed on a cyclical basis at intervals not to exceed 30 minutes.~~ A system or portion of a combination system that utilizes one or more stationary sensors to detect the presence of a specified gas at a specified concentration and initiate one or more responses required by this code, such as notifying a responsible person, activating an alarm signal, or activating or deactivating equipment. A self-contained gas detection and alarm device is not classified as a gas detection system.

CHANGE SIGNIFICANCE: Gas detection systems are required for many different applications in the code. These revisions correlate all of these requirements and provide consistency in their application. Additionally, several changes occurred since some of the requirements in the 2015 IFC cannot be enforced because listed gas detectors, controls and systems are not commercially available.

Several significant changes occurred as follows:

- Section 105.7.11 – this section is added, which requires a construction permit for installation of gas detection systems.
- Section 202 – the definition of "Gas detection system" is deleted. The requirement for continuous operation is removed from the definition and addressed in Section 916.7.
- Sections 916.5 and 1203.2.6 – gas detection systems are required to be provided with emergency or standby power. By default, Section 1203.2.6 requires a minimum duration of 2 hours. An option is included for gas detection systems to provide a power loss trouble signal in an approved location in lieu of standby power.

Gas detection systems are required in the following sections:

- Section 1205.13 for fuel cell power systems
- Section 1206.2.12.6 for battery technologies not specifically regulated in the code
- Section 2311.8.8.2 as an option for mechanical exhaust ventilation in repair garages for lighter-than-air-fueled vehicles
- Section 2311.8.9 for repair garages for lighter-than-air-fueled vehicles which are nonodorized
- Section 2703.13 for certain HPM gases in semiconductor fabrication facilities
- Section 3905.1 for processing and extraction facilities
- Section 5307.2.1 as an option to continuous ventilation for compressed gases which represent asphyxiant, irritant or radioactive hazards
- Section 5307.3.2 as an option to continuous ventilation for liquid carbon dioxide systems for beverage dispensing
- Section 5307.4.3 for carbon dioxide enrichment systems
- Section 5808.5 for hydrogen fuel gas rooms
- Section 6004.2.2.10 for toxic and highly toxic gases located indoors
- Section 6004.3.4.3 for toxic and highly toxic gases located outdoors
- Section 6005.3.2 for ozone gas generator rooms

In the 2018 IFC, the sections above reference Section 916 for gas detection system design. Note that mechanical refrigeration systems are not listed. That is intentional because Section 605.8 requires ammonia refrigerant systems to comply with IIAR 2 "Safe Design of Closed-Circuit Ammonia Refrigerating Systems."

The revised definition for gas detection systems specifies that a standalone device, which detects gas and sounds an alarm, is not considered to be a gas detection system. This is similar to the concept of smoke detection systems vs. smoke alarms.

Section 916 includes basic requirements for all gas detection systems and covers construction documents, equipment, power connections, emergency and standby power, sensor locations, gas sampling, system activation, signage, fire alarm system connections, and maintenance. The gas detection system must be inspected and tested at the time of installation, and Section 916.11 specifies that annual inspection and testing is required. Sensor calibration is also required at the time of installation, with recalibration occurring at frequencies required by the manufacturer.

Table 1004.5, 1004.8

Occupant Load Calculation in Business Use Areas

CHANGE TYPE: Modification

CHANGE SUMMARY: The method of calculating occupant load in business areas is revised, which allows for larger occupant loads.

2015 CODE:

TABLE ~~1004.1.2~~ 1004.5 Maximum Floor Area Allowances per Occupant

FUNCTION OF SPACE	OCCUPANT LOAD FACTOR[a]
Business areas	~~100~~ 150 gross
Concentrated business use areas	See Section 1004.8

For SI: 1 square foot = 0.0929 m^2, 1 foot = 304.8 mm.
a. Floor area in square feet per occupant.

1004.8 Concentrated business use areas. The occupant load factor for concentrated business use shall be applied to telephone call centers, trading floors, electronic data processing centers and similar business use areas with a higher density of occupants than would normally be expected in a typical business occupancy environment. Where approved by the code official, the occupant load for concentrated business use areas shall be the actual occupant load, but not less than one occupant per 50 square feet (4.65 m^2) of gross occupiable floor space.

CHANGE SIGNIFICANCE: This code change makes two changes to calculating the occupant load in a business use area.

The first change revises the 100 square feet per occupant to 150 square feet per occupant. for determining the means of egress requirements in business areas. The occupant load factor of 100 square feet per occupant for business occupancies first appeared in the 3rd edition of the *Building Exits Code*, published in 1934. The occupant load factor of 100 square feet

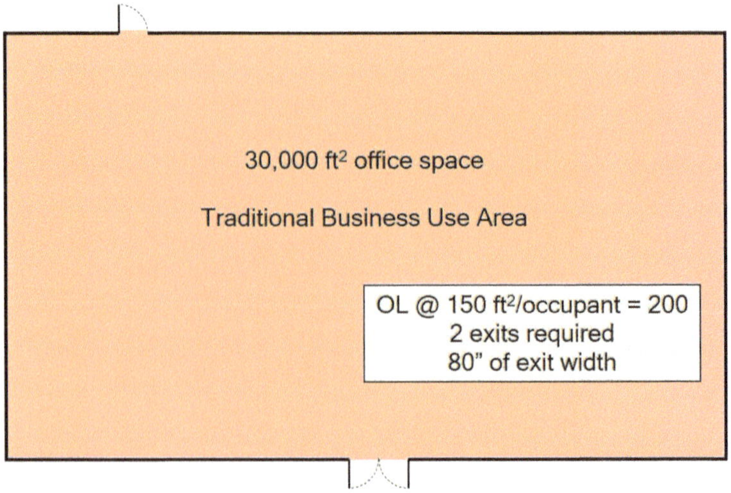

Typical business use is based on an occupant load factor of 150 square feet per occupant.

per occupant was specified for office, factory, and workroom areas. All occupant load factors were based on the gross floor area of the building, such that no deduction was permitted for corridors, closets, restrooms, or other subdivisions. It seems likely that the results from a National Bureau of Standards (NBS) [now referred to as National Institute of Standards and Technology (NIST)] study published in 1935 were the most likely basis of the occupant load factors adopted into the 1934 Code. However, since the initial NBS study in 1935, several other studies have been conducted to determine occupant load factors for various occupancies. One common similarity of each of the studies was that all of the subsequent studies have concluded that the factor of 100 square feet per occupant for business occupancies is conservative. Studies conducted between 1966 and 1992 have indicated that occupant load factors in business occupancies ranged from 150 to 278 square feet per occupant. In addition, a 1995 study of 23 Federal sector and private sector office buildings also indicated a mean occupant load factor of 248 square feet per occupant for all office buildings. A more recent project to study the appropriateness of the 100 square feet per occupant factor was undertaken by the NFPA Fire Protection Research Foundation (NFPRF). The study was conducted by Worchester Polytechnic Institute undergrad students. The recommendations of this study have indicated that it is reasonable to increase the occupant load factor to square feet per occupant in business occupancies.

The NFPRF study also recommended creating a new occupant load subcategory for concentrated business use areas. This has resulted in the second change—the allowance to use an increased occupant load factor for high occupancy areas in business occupancies. Section 1004.8 cites examples of these occupancies as telephone call centers, trading floors, and electronic data processing centers; essentially, business areas where a higher density of occupants would normally be expected.

This increased occupant load factor cannot go beyond 50 square feet per occupant, and is only allowed where approved by the code official.

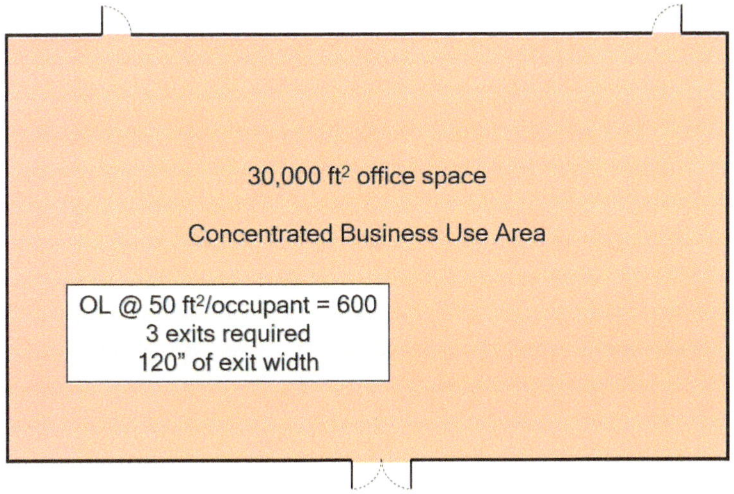

Concentrated business use areas can based on an occupant load factor of 50 square feet per occupant.

1006.2.1

Spaces with One Exit or Exit Access Doorway

CHANGE TYPE: Modification

CHANGE SUMMARY: Determination of cumulative occupant loads is clarified and correlated with other code requirements.

2018 CODE: 1006.2.1 Egress based on occupant load and common path of egress travel distance. Two *exits* or *exit access doorways* from any space shall be provided where the design *occupant load* or the *common path of egress travel* distance exceeds the values listed in Table 1006.2.1. The cumulative occupant load from adjacent rooms, areas or spaces shall be determined in accordance with Section 1004.2.

Exceptions:

~~1. In Group R-2 and R-3 occupancies, one means of egress is permitted within and from individual dwelling units with a maximum occupant load of 20 where the dwelling unit is equipped throughout with an automatic sprinkler system in accordance with Section 903.3.1.1 or 903.3.1.2 and the common path of egress travel does not exceed 125 feet (38 100 mm).~~

1. The number of exits from foyers, lobbies, vestibules or similar spaces need not be based on cumulative occupant loads for areas discharging through such spaces, but the capacity of the exits from such spaces shall be based on applicable cumulative occupant loads.
2. Care suites in Group I-2 occupancies complying with Section 407.4 of the *International Building Code*.

TABLE 1006.2.1 *(PARTIAL)* Spaces with One Exit or Exit Access Doorway

OCCUPANCY	MAXIMUM OCCUPANT LOAD OF SPACE	MAXIMUM COMMON PATH OF EGRESS TRAVEL DISTANCE (feet)		
		Without Sprinkler System (feet)		With Sprinkler System (feet)
		Occupant Load		
		OL ≤ 30	OL >30	
R-2	~~10~~20	NP	NP	125[a]
R-3[e]	~~10~~20	NP	NP	125[a, g]
R-4[e]	~~10~~20	~~75~~NP	~~75~~NP	125[a, g]

For SI: 1 foot = 304.8 mm.
NP = Not Permitted
a. *No change in text...*
b. *No change in text...*
c. *No change in text...*
d. *No change in text...*
e. The ~~length of~~ common path of egress travel distance shall apply only in a Group R-3 occupancy located in a mixed occupancy building or within a Group R-3 or R-4 congregate living facility.
f. *No change in text...*
g. For the travel distance limitations in Groups R-3 and R-4 equipped throughout with an automatic sprinkler system in accordance with Section 903.3.1.3, see Section 1006.2.2.6.

CHANGE SIGNIFICANCE: The revision in Section 1006.2.1 correlates this section with Section 1004.2.1. Section 1004.2.1 states, in part: "Design of egress path capacity shall be based on the cumulative portion of the occupant loads of all rooms, areas or spaces to that point along the path of egress travel."

This proposal places the same requirement at Section 1006.2.1 in the context of using the cumulative capacity to determine the required number of exits or access to exits.

The previous Exception 1 to Section 1006.2.1 is deleted. This exception stated that the maximum occupant load for one exit in Group R-2 or R-3 is 20 if the building is protected with an automatic sprinkler system. Since all new Group R-2 and R-3 are required to be sprinklered, this exception is always applicable. To correlate with this change, the entries in Table 1006.2.1 have been revised to acknowledge the occupant load of 20 when one exit is provided. Group R-4 is also revised to an allowed occupant load of 20 since Group R-4 occupancies are limited to an occupant load of 16 not including staff. Therefore, it is not likely that the occupant load will exceed 20. See also Significant Change to Section 1006.2.2.6.

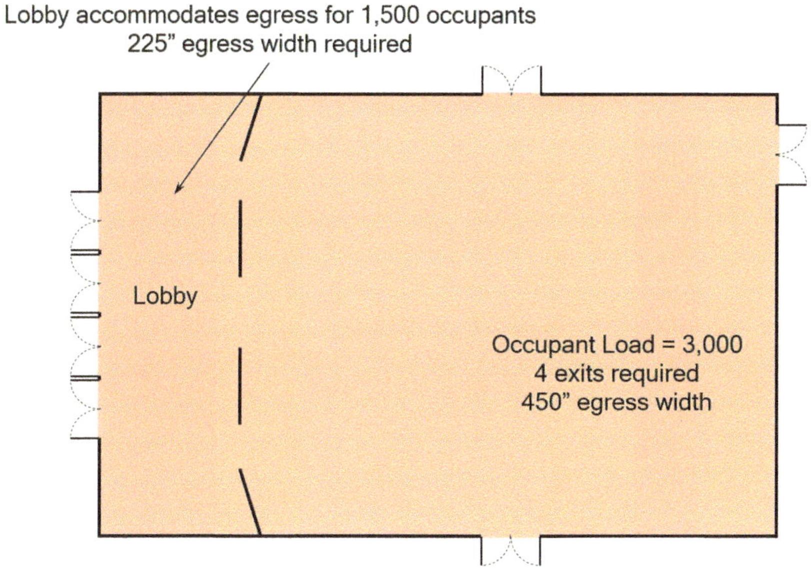

Cumulative occupant load egressing through lobby is used to determine required egress width, but is not used to determine the number of exits from the lobby.

New Exception 1 is added to address the situation where a lobby or foyer becomes the intervening room. Literally interpreted, a building with an occupant load of 3,000 and having four required exits with one of those exits traveling through a lobby or foyer would then also require four exits from such lobby based on the cumulative occupant load of 1,000. To resolve that, Exception 1 states that the number of exits from such space is not to be based on the cumulative occupant load; however, the capacity, or egress width, of the exits is based on the cumulative occupant load served.

1006.2.1 continues

1006.2.1 continued

The main exit must still accommodate at least one-half of the required exit width when an assembly has an occupant load greater than 300. Section 1029.2 correlates with the requirements in Section 1006.2.1 by addressing the egress width from the lobby rather than recalculating the number of exits from the lobby.

The addition of Footnote g to Table 1006.2.1 correlates with revisions to Table 1017.2. See Significant Change to Table 1017.2.

1006.2.2.6, 1006.2.1, 1017.2

Groups R-3 and R-4 Protected with NFPA 13D Sprinkler System

CHANGE TYPE: Addition

CHANGE SUMMARY: Exit access travel distances are provided for Groups R-3 and R-4 when sprinklered with NFPA 13D sprinkler systems.

2018 CODE: **1006.2.2.6 Groups R-3 and R-4.** Where Group R-3 occupancies are permitted by Section 903.2.8 to be protected by an automatic sprinkler system installed in accordance with Section 903.3.1.3, the exit access travel distance for Group R-3 shall be not more than 125 feet (38 100 mm). Where Group R-4 occupancies are permitted by Section 903.2.8 to be protected by an automatic sprinkler system installed in accordance with Section 903.3.1.3, the exit access travel distance for Group R-4 shall be not more than 75 feet (22 860 mm).

1006.2.2.6, 1006.2.1, 1017.2 continues

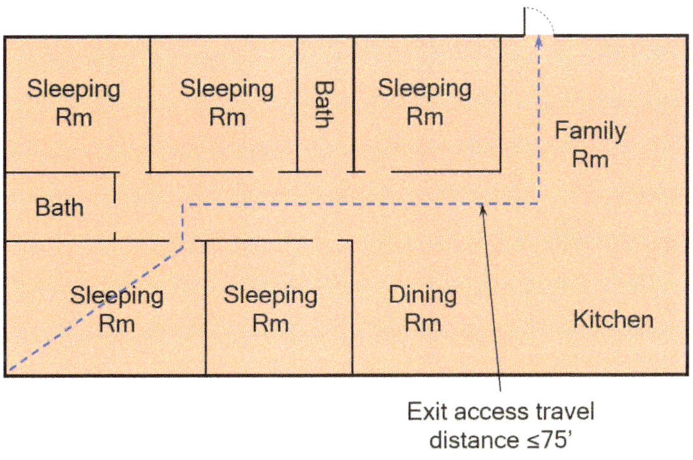

The exit access travel distance is limited to 75 feet in Group R-4 occupancies equipped with sprinkler systems designed to NFPA 13D.

1006.2.2.6, 1006.2.1, 1017.2
continued

TABLE 1006.2.1 *(PARTIAL)* **Spaces with One Exit or Exit Access Doorway**

OCCUPANCY	MAXIMUM OCCUPANT LOAD OF SPACE	MAXIMUM COMMON PATH OF EGRESS TRAVEL DISTANCE (feet)		
		Without Sprinkler System (feet)		With Sprinkler System (feet)
		Occupant Load		
		OL ≤ 30	OL > 30	
R-1	10	NP	NP	75[a]
R-2	~~10~~20	NP	NP	125[a]
R-3[e]	~~10~~20	NP	NP	125[a, g]
R-4[e]	~~10~~20	~~75~~NP	~~75~~NP	125[a, g]

For SI: 1 foot = 304.8 mm.
NP = Not Permitted
a. *No change in text...*
b. *No change in text...*
c. *No change in text...*
d. *No change in text...*
e. The ~~length of~~ common path of egress travel distance shall apply only in a Group R-3 occupancy located in a mixed occupancy building or within a Group R-3 or R-4 congregate living facility.
f. *No change in text...*
g. For the travel distance limitations in Groups R-3 and R-4 equipped throughout with an automatic sprinkler system in accordance with Section 903.3.1.3, see Section 1006.2.2.6.

TABLE 1017.2 *(partial)* **Exit Access Travel Distance[a]**

OCCUPANCY	WITHOUT SPRINKLER SYTEM (feet)	WITH SPRINKLER SYSTEM (feet)
A, E, F-1, M, R, S-1	200	250[b, e]

For SI: 1 foot = 304.8 mm.
a. *No change in text...*
b. Buildings equipped throughout with an automatic sprinkler system in accordance with Section 903.3.1.1 or 903.3.1.2. See Section 903 for occupancies where automatic sprinkler systems are permitted in accordance with Section 903.3.1.2.
c. *No change in text...*
d. *No change in text...*
e. Group R-3 and R-4 buildings equipped throughout with an automatic sprinkler system in accordance with Section 903.3.1.3. See Section 903.2.8 for occupancies where automatic sprinkler systems are permitted in accordance with Section 903.3.1.3.

CHANGE SIGNIFICANCE: New Section 1006.2.2.6 is added that changes the exit access travel distance to 125 feet in Group R-3 occupancies protected with a sprinkler system designed to NFPA 13D, "Standard for the Installation of Sprinkler Systems in One- and Two-Family Dwellings and Manufactured Homes," and to 75 feet in Group R-4 occupancies protected with a sprinkler system designed to NFPA 13D. This section is referenced from Table 1006.2.1 in Footnote g.

Looking at Footnote g in Table 1006.2.1, notice that Footnote e has been clarified. The common path of travel limitations in the table only apply to Group R-3 in a mixed occupancy building, Group R-3 congregate living facilities, or Group R-4 congregate living facilities.

Additionally, Footnote g has been added which refers to Section 1006.2.1 for Groups R-3 and R-4 sprinklered with NFPA 13D sprinkler systems. The resulting distances for common path of egress travel are shown in Table 1.

TABLE 1 Common Path of Egress Travel Distance

GROUP	DESIGN	NOT SPRINKLERED	SPRINKLERED WITH NFPA 13	SPRINKLERED WITH NFPA 13R	SPRINKLERED WITH NFPA 13D
R-3	In a Mixed Occupancy Building	Not Permitted	125 feet	125 feet	125 feet
	Congregate Living Facility	Not Permitted	125 feet	125 feet	125 feet
	All other situations	Not Permitted	Not Limited	Not Limited	125 feet
R-4	In a Mixed Occupancy Building	Not Permitted	125 feet	125 feet	75 feet
	Congregate Living Facility	Not Permitted	125 feet	125 feet	75 feet
	All other situations	Not Permitted	Not Limited	Not Limited	75 feet

The common path of travel distance in Table 1006.2.1 for Group R-4 shows 125 feet, however, Table 1 above shows the distance as 75 feet. This limitation to 75 feet occurs because Section 1006.2.2.6 limits the total exit access travel distance to 75 feet. The common path of egress travel cannot exceed the allowed exit access travel distance.

For a discussion on the revision from an occupant load of 10 to 20, see Significant Change to Section 1006.2.1.

Table 1017.2 is also revised for Groups R-3 and R-4. Footnote e is added which correlates with Footnote b. Footnote b indicates that the exit access travel distance for Groups R-3 and R-4 is 250 feet when sprinklered with NFPA 13 or NFPA 13D. Footnote e clarifies that for Groups R-3 or R-4 sprinklered with NFPA 13D, the exit access travel distance is limited to 200 feet. The resulting exit access travel distances are shown in Table 2.

TABLE 2 Exit Access Travel Distance

GROUP	DESIGN	NOT SPRINKLERED	SPRINKLERED WITH NFPA 13	SPRINKLERED WITH NFPA 13R	SPRINKLERED WITH NFPA 13D
R-3	In a Mixed Occupancy Building	200 feet	250 feet	250 feet	125 feet
	Congregate Living Facility	200 feet	250 feet	250 feet	125 feet
	All other situations	200 feet	250 feet	250 feet	200 feet
R-4	In a Mixed Occupancy Building	200 feet	250 feet	250 feet	75 feet
	Congregate Living Facility	200 feet	250 feet	250 feet	75 feet
	All other situations	200 feet	250 feet	250 feet	75 feet

1006.3, 1006.3.1

Exits on Adjacent Stories

CHANGE TYPE: Modification

CHANGE SUMMARY: Determining egress requirements has been clarified when the occupants travel to an adjacent story to reach the exit.

2018 CODE: **1006.3 Egress from stories or occupied roofs.** The *means of egress* system serving any *story* or occupied roof shall be provided with the number of separate and distinct *exits* or access to *exits* based on the aggregate *occupant load* served in accordance with this section. ~~The path of egress travel to an exit shall not pass through more than one adjacent story.~~ Where stairways serve more than one story, only the occupant load of each story considered individually shall be used in calculating the required number of exits or access to exits serving that story.

1006.3.1 Adjacent story. The path of egress travel to an exit shall not pass through more than one adjacent story.

> **Exception:** The path of egress travel to an *exit* shall be permitted to pass through more than one adjacent *story* in any of the following:
>
> 1. In Group R-1, R-2 or R-3 occupancies, exit access stairways and ramps connecting four stories or fewer serving and contained within an individual dwelling unit or sleeping unit or live/work unit.
> 2. Exit access stairways serving and contained within a Group R-3 congregate residence or a Group R-4 facility.
> 3. Exit access stairways and ramps in open parking garages that serve only the parking garage.
> 4. Exit access stairways and ramps serving open-air assembly seating complying with the exit access travel distance requirements of Section 1029.7.
> 5. Exit access stairways and ramps between the balcony, gallery or press box and the main assembly floor in occupancies such as theaters, places of religious worship, auditoriums and sports facilities.

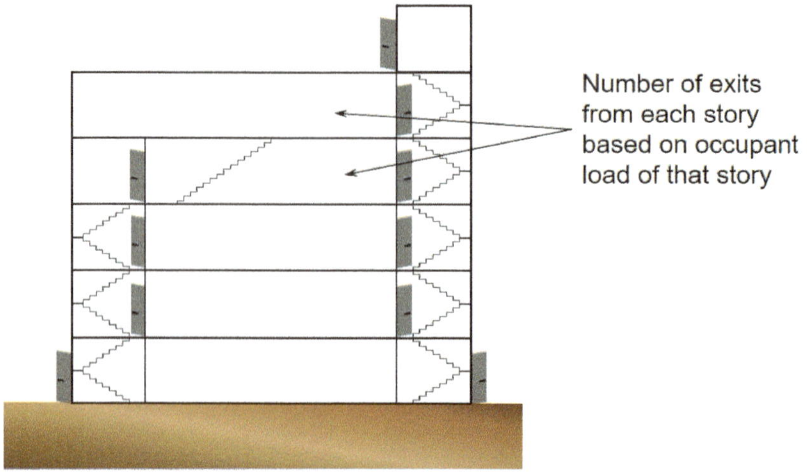

The number of exits, or access to exits, from each story is determined based on the occupant load for the that story alone. The additional occupant load coming from the adjacent story is not added.

CHANGE SIGNIFICANCE: In the 2015 IFC, the second sentence of Section 1006.3 states that the required number of exits must be available not more than one story above or below the floor level you are on. That sentence is relocated to Section 1006.3.1 and several exceptions are added to clarify application of this allowance. The exceptions correlate with Section 1019.3 which already allows for unenclosed exit access stairways to be provided in these situations. The requirements in Sections 1019.3 and 1006.3.1 now work together to allow the egress path to traverse more than one story to reach the exit and for the use of unenclosed stairways for those situations.

The last sentence added into Section 1006.3 now states that in situations where the occupants egress through an adjacent story to reach the exit, the additional occupant load is not considered when determining the required number of exits from the adjacent story. The end result is that the number of exits, or access to exits, for each story is based on the occupant load for that story.

Section 1004.2.3 also states that the occupant load from the adjacent story is not added when determining the required egress width. Therefore, the number of exits, or access to exits, and the required egress width is based on the occupant load of each story. Keep in mind that this is specific to an adjacent story, and is treated differently than a mezzanine. The occupant load from a mezzanine is added when calculating the required exits and egress width.

1008.2.3

Illumination of the Exit Discharge

CHANGE TYPE: Clarification

CHANGE SUMMARY: Illumination of exit discharge can now terminate at a safe dispersal area.

2018 CODE: **1008.2.3 Exit discharge.** Illumination shall be provided along the path of travel for the exit discharge from each exit to the public way.

> **Exception:** Illumination shall not be required where the path of the exit discharge meets both of the following requirements:
>
> 1. The path of exit discharge is illuminated from the exit to a safe dispersal area complying with Section 1028.5.
> 2. A dispersal area shall be illuminated to a level not less than 1 footcandle (11 lux) at the walking surface.

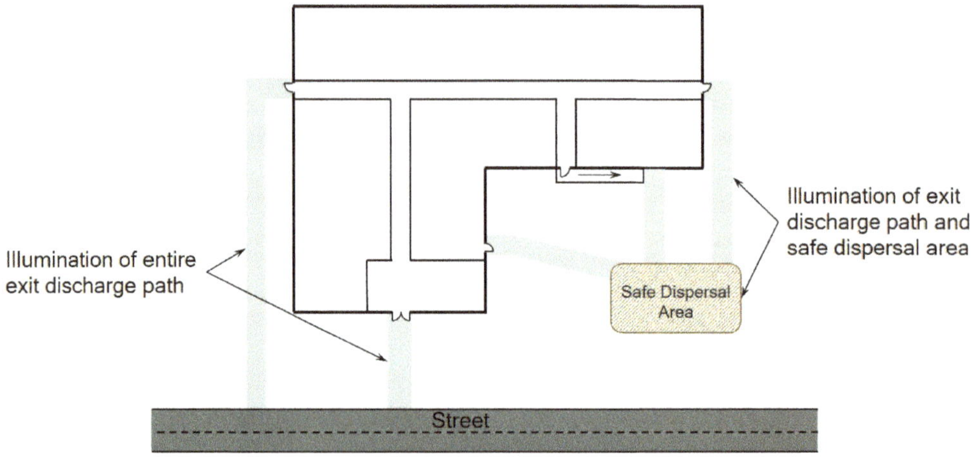

Illumination of the exit discharge is required either to the public way, or to a safe dispersal area.

CHANGE SIGNIFICANCE: This change limits the amount of light required for safe exiting from a building. On large parcels and when buildings are constructed on existing private campuses (i.e., business parks, college/university), the need to provide required lighting to the public way can be significant.

This section specifies that illumination must be provided for the entire exit discharge path to the public way, or to a safe dispersal area. By design, in Section 1028.5, a safe dispersal area is located at least 50 feet from the building and provides adequate area to accommodate the anticipated occupant load. This section requires illumination to, and within, the safe dispersal area. This new section addresses the normal illumination levels and states a minimum of 1 footcandle must be provided at the walking surface.

As required in Section 1028.5, the safe dispersal must be identified and maintained. It cannot be just somewhere out in the parking lot or in the grassy field. It must provide at least 5 square feet per person and at least 50 from the building where the occupants just exited.

Note that this section does not require emergency power or emergency illumination for the entire exit discharge path, or even to the safe dispersal area. Emergency illumination is only required at exterior landings at exit doors as stated in Section 1008.3.2.

See also the Significant Change to Section 1008.2.2 for required illumination levels in Group I-2.

1008.3.5, 1008.2.2

Emergency Illumination in Group I-2

CHANGE TYPE: Modification

CHANGE SUMMARY: Emergency egress lighting in Group I-2 must meet minimum illumination levels even when one lamp fails in a single luminaire.

2018 CODE: **1008.3.5 Illumination level under emergency power.** Emergency lighting facilities shall be arranged to provide initial illumination that is not less than an average of 1 footcandle (11 lux) and a minimum at any point of 0.1 footcandle (1 lux) measured along the path of egress at floor level. Illumination levels shall be permitted to decline to 0.6 footcandle (6 lux) average and a minimum at any point of 0.06 footcandle (0.6 lux) at the end of the emergency lighting time duration. A maximum-to-minimum illumination uniformity ratio of 40 to 1 shall not be exceeded. In Group I-2 occupancies, failure of ~~any~~ a single ~~lighting unit~~ lamp in a luminaire shall not reduce the illumination level to less than 0.2 foot-candle (2.2 lux).

Emergency egress lighting in Group I-2 must maintain a minimum lighting level when one lamp, or bulb, fails in the emergency lighting units.

1008.2.2 ~~Exit discharge~~ Group I-2. In Group I-2 occupancies where two or more *exits* are required, on the exterior landings required by Section 1010.6.1, means of egress illumination levels for the *exit discharge* shall be provided such that failure of ~~any~~ a single ~~lighting unit~~ lamp in a luminaire shall not reduce the illumination level on that landing to less than 1 footcandle (11 lux).

CHANGE SIGNIFICANCE: These revisions change the requirement for emergency lighting in Group I-2 occupancies. Previous language required that the illumination level must be provided when a single luminaire fails. Article 100 of the 2017 *National Electrical Code*® defines a luminaire as a "...lighting unit consisting of a lamp or lamps...". Therefore, failure of any single lighting unit was considered in the 2015 IFC.

This revised section states that the illumination must be provided with the failure of one single lamp, rather than the entire lighting unit. So, the evaluation of the emergency lighting system will consider one of the lamps has failed, rather than both lamps in a luminaire with two lamps.

The evaluation of a loss of single lamp is only applicable to Group I-2 occupancies. The illumination level for egress illumination must be maintained at 0.2 footcandle with the loss of one lamp; and illumination of exterior landings at 1 footcandle.

See also the Significant Change to Section 1008.2.3 for illumination of exit discharge.

1009.7.2

Protection of Exterior Areas of Assisted Rescue

CHANGE TYPE: Modification

CHANGE SUMMARY: The 1-hour fire-resistance-rated separation between an exterior of assisted rescue and the building is not required if the building is protected with an automatic sprinkler system designed to NFPA 13 or 13R.

2018 CODE: **1009.7.2 Separation.** *Exterior walls* separating the exterior area of assisted rescue from the interior of the building shall have a minimum fire-resistance rating of 1 hour, rated for exposure to fire from the inside. The fire-resistance-rated *exterior wall* construction shall extend horizontally not less than 10 feet (3048 mm) beyond the landing on either side of the landing or equivalent fire-resistance-rated construction is permitted to extend out perpendicular to the *exterior wall* not less than 4 feet (1220 mm) ~~minimum~~ on the side of the landing. The fire-resistance-rated construction shall extend vertically from the ground to a point not less than 10 feet (3048 mm) above the floor level of the area for assisted rescue or to the roof line, whichever is lower. Openings within such fire-resistance-rated *exterior walls* shall be protected in accordance with Section 716 of the *International Building Code.*

> **Exception:** The fire-resistance rating and opening protectives are not required in the exterior wall where the building is equipped throughout with an automatic sprinkler system installed in accordance with Section 903.3.1.1 or 903.3.1.2.

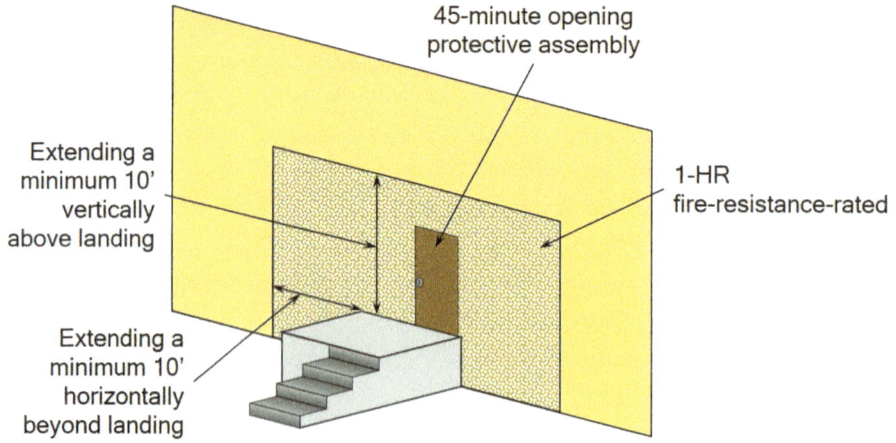

The exterior area of assisted rescue is protected by either fire-resistance-rated construction or the building is sprinklered to NFPA 13 or 13R.

CHANGE SIGNIFICANCE: Exterior areas of assisted rescue can be provided when the exit discharge path does not consist of an accessible route completely to the public way. The exterior area of assisted rescue provides a location for the mobility impaired person to wait for assistance to arrive.

The area of assisted rescue is to be protected from the interior of the building by a 1-hour fire-resistance-rated wall with opening protectives unless the building is equipped with an automatic sprinkler system designed to NFPA 13, "Standard for Installation of Sprinkler Systems" or 13R, "Standard for Installation of Sprinkler Systems in Low-Rise Residential Occupancies."

Section 1009.3.3, Exception 2 already allows elimination of areas of refuge in stairways when the building is sprinklered. This new exception to Section 1009.7.2 carries this concept one step further. The concept is that if the person was adequately protected inside the building because it was sprinklered, now that they are outside the building the level should not be any worse. So, if the building is sprinklered, the fire-resistance-rated separation is not required.

This exception only allows the elimination of the fire-resistance-rated construction. It does not eliminate the need to provide the exterior area of assisted rescue. The exterior area of assisted rescue must still be sized to provide one wheelchair space for every 200 occupants, or portion thereof. It must still be at least 50 percent open to the outside air. If the exterior area for assisted rescue is provided with stairs, the stairs do not need to meet the minimum clear width of 48 inches measured between the handrails, since the building is sprinklered.

1010.1.1

Size of Doors

CHANGE TYPE: Clarification

CHANGE SUMMARY: Requirements for the size of doors is revised to correlate with ICC A117.1.

2018 CODE: 1010.1.1 Size of doors. The required capacity of each door opening shall be sufficient for the *occupant load* thereof and shall provide a minimum clear opening width of 32 inches (813 mm). ~~Clear openings~~ The clear opening width of doorways with swinging doors shall be measured between the face of the door and the stop, with the door open 90 degrees (1.57 rad). Where this section requires a minimum clear opening width of 32 inches (813 mm) and a door opening includes two door leaves without a mullion, one leaf shall provide a minimum clear opening width of 32 inches (813 mm). In Group I-2, doors serving as means of egress doors where used for the movement of beds shall provide a minimum clear opening width of 41½ inches (1054 mm). The maximum width of a swinging door leaf shall be 48 inches (1219 mm) nominal. ~~Means of egress doors in a Group I-2 occupancy used for the movement of beds shall provide a clear width not less than 41½ inches (1054 mm).~~ The minimum clear opening height of ~~door openings~~ doors shall be not less than 80 inches (2032 mm).

Exceptions:

1. ~~The~~ In Group R-2 and R-3 dwelling and sleeping units that are not required to be an Accessible, Type A unit or Type B unit, the minimum and maximum width shall not apply to door openings that are not part of the required *means of egress*. ~~in Group R-2 and R-3 occupancies.~~
2. ~~Door~~ Group I-3 door openings to resident *sleeping units* ~~in Group I-3 occupancies~~ that are not required to be an Accessible unit shall have a minimum clear opening width of ~~not less than~~ 28 inches (711 mm).
3. Door openings to storage closets less than 10 square feet (0.93 m^2) in area shall not be limited by the minimum clear opening width.
4. ~~Width~~ The width of door leaves in revolving doors that comply with Section 1010.1.4.1 shall not be limited.
5. The maximum width of door leaves in power-operated doors that comply with Section 1010.1.4.2 shall not be limited.
6. Door openings within a *dwelling unit* or *sleeping unit* shall ~~be not less than~~ have a minimum clear opening height of 78 inches (1981 mm). ~~in height.~~
~~6.~~7. ~~Exterior door openings in dwelling units and sleeping units, other than the required exit door, shall be not less than 76 inches (1930 mm) in height.~~ In dwelling and sleeping units that are not required to be Accessible, Type A or Type B units, exterior door openings, other than the required *exit* door, shall have a minimum clear opening height of 76 inches (1930 mm).

~~7.~~8. ~~In other than Group R-1 occupancies, the minimum widths shall not apply to interior egress doors within a dwelling unit or sleeping unit that is not required to be an Accessible unit, Type A unit or Type B unit.~~ In Groups I-1, R-2, R-3 and R-4, in dwelling and sleeping units that are not required to be Accessible, Type A or Type B units, the minimum clear opening widths shall not apply to interior egress doors.

~~8.~~9. Door openings required to be *accessible* within Type B units intended for user passage shall have a minimum clear opening width of 31.75 inches (806 mm).

~~9.~~10. Doors to walk-in freezers and coolers less than 1,000 square feet (93 m^2) in area shall have a maximum width of 60 inches (1524 mm) nominal.

~~10.~~11. ~~In Group R-1 dwelling units or sleeping units not required to be Accessible units, the~~ The minimum clear opening width shall not apply to doors for nonaccessible shower or sauna compartments.

12. The minimum clear opening width shall not apply to the doors for nonaccessible toilet stalls.

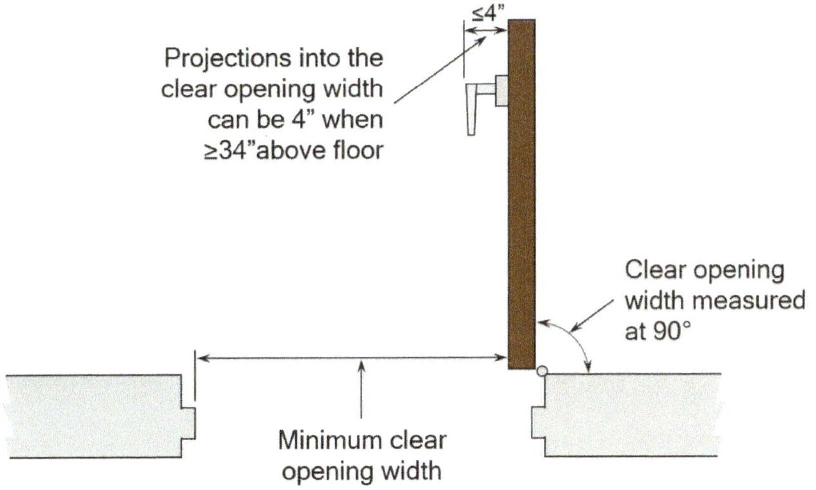

The clear opening width is measured with the door open to 90 degrees. The required minimum clear opening width varies based on the occupancy.

CHANGE SIGNIFICANCE: This is a reformat of the text which provides consistency throughout the section. The same term is used in each case (e.g., minimum clear opening width and height). Also, the modifier appears first within each specific requirement (Group I-1, Group R-2, etc.).

A reference to nonaccessible units is added to Exceptions 1, 2, 7 and 8. This provides consistency with ICC A117.1 "Accessible and Usable Buildings and Facilities," ADA Standards, and FHA. Exception 8 is further modified to recognize that dwelling units and sleeping units in Group I-2 and I-3 have specific criteria elsewhere in this section, and the ADA does not allow Group R-1 units to use this exception. Therefore, this exception is limited to Groups I-1, R-2, R-3, and R-4.

1010.1.1 continues

1010.1.1 continued

Exception 9 has been revised to be consistent with the language in ICC A117.1 for Type B dwelling units.

Exception 11 is revised to state that the exception is not limited to nonaccessible Group R-1 dwelling units, but rather it more specifically applies to all showers and sauna compartments that are not required to be accessible.

Exception 12 is added to address a similar question for doors on toilet stalls. The width of 32 inches is especially a problem with the *International Plumbing Code*® since the stall is only required to be 30 inches wide.

1010.1.4.4

Locking Arrangements in Educational Occupancies

CHANGE TYPE: Addition

CHANGE SUMMARY: Guidance is provided to allow enhanced security measures yet still meet egress requirements for classroom doors.

2018 CODE: **1010.1.4.4 Locking arrangements in educational occupancies.** In Group E and Group B educational occupancies, egress doors from classrooms, offices and other occupied rooms shall be permitted to be provided with locking arrangements designed to keep intruders from entering the room, where all of the following conditions are met:

1. The door shall be capable of being unlocked from outside the room with a key or other approved means.
2. The door shall be openable from within the room in accordance with Section 1010.1.9.
3. Modifications shall not be made to listed panic hardware, fire door hardware or door closers.

1010.1.4.4.1 Remote operation of locks. Remote operation of locks complying with Section 1010.1.4.4 shall be permitted.

Photo courtesy of Schlage Lock Co., LLC (part of Allegion plc)

This electrically operated lock is openable from outside the classroom by entering the proper pass code.

CHANGE SIGNIFICANCE: A high priority in many locations is the safety of occupants at schools and in classrooms during the event of a threatening situation. These revisions provide criteria in the code which balance the challenges of providing protection for students and teachers in the classroom while at the same time providing for free and immediate egress.

Oftentimes locks or devices are added to doors to provide security, but they do not allow free egress, or comply with the single operation requirement. These provisions provide guidance for combining security while maintaining safe egress.

Door locksets with "classroom security function" are readily available at a comparable cost to the traditional "classroom function" door locksets. The most common configuration of a classroom security function lockset is the ability to lock the door from inside the classroom with a key preventing entry into the classroom; and for egress, the door may be easily opened from inside the classroom without a key by a single action on the lever handle. On the outside of the classroom, consistent with tradition, the door may be locked with a key, and unlocked and opened with a key.

Many of the traditional locksets required the instructor to leave the classroom and lock the door with a key from the hallway side, then re-enter the classroom for a defend-in-place strategy. This action places the instructor at risk by forcing them to leave the classroom and become exposed in the hallway. The classroom security function eliminates the need to leave the classroom to lock the door and it still allows unrestricted egress from inside the classroom.

1010.1.4.4 continues

1010.1.4.4 continued

Section 1010.4.4.1 is added which allows the remote operation of classroom locks. This means that the doors can be locked electrically from a single location. This design allows all of the classroom doors to be locked, and unlocked, with the operation of a single switch. This switch can be designed to be key operated to add an additional level of security.

Additionally, this language requires that the door shall be unlockable from outside the classroom. This allows fire fighters, school personnel, or law enforcement to obtain entry even after the door is locked from the inside. This can be accomplished at the door with a key, or remotely with electric lock operation.

See also Significant Change to Section 1031.2.2 for locking arrangements in existing Group I-4 day care facilities.

1010.1.9.8, 1010.1.9.8.1

Delayed Egress

CHANGE TYPE: Modification

CHANGE SUMMARY: Additional occupancies are allowed to install delayed egress, including small Group E occupancies and Group A courtrooms.

2018 CODE: **1010.1.9.8 Delayed egress.** Delayed egress locking systems shall be permitted to be installed on doors serving ~~any occupancy except Group A, E and H~~ the following occupancies in buildings that are equipped throughout with an *automatic sprinkler system* in accordance with Section 903.3.1.1 or an *approved* automatic smoke or heat detection system installed in accordance with Section 907:

1. Group B, F, I, M, R, S and U occupancies.

2. Group E classrooms with an occupant load of less than 50.

Exception: Delayed egress locking systems shall be permitted to be installed on exit or exit access doors, other than the main exit or exit access door, serving a courtroom in buildings equipped throughout with an *automatic sprinkler system* in accordance with Section 903.3.1.1.

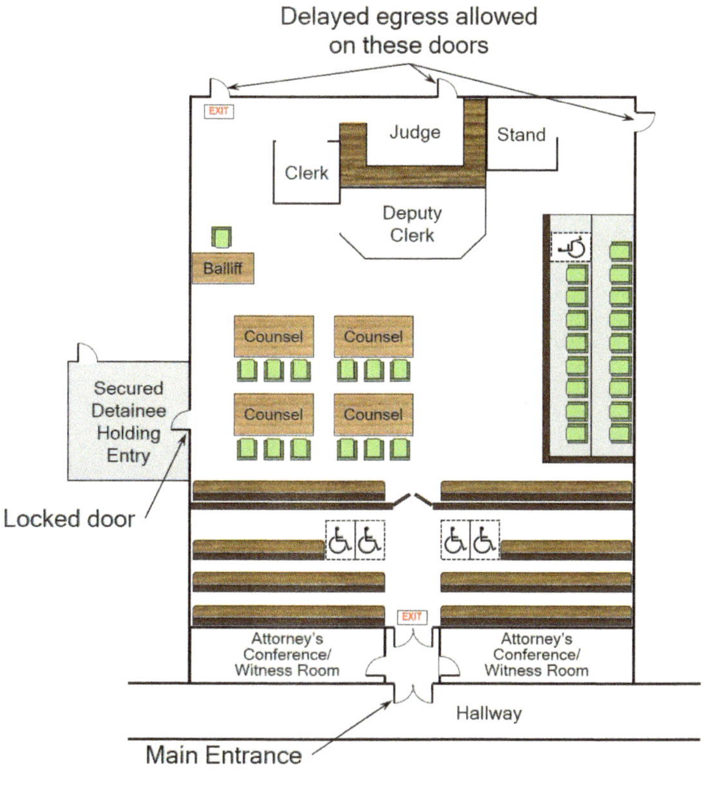

The main exit in the courtroom provides for free and uninhibited egress. Additional exit doors or exit access doors can be provided with delayed egress.

1010.1.9.8, 1010.1.9.8.1 continues

1010.1.9.8, 1010.1.9.8.1 continued

1010.1.9.8.1 Delayed egress locking system. The delayed egress locking system shall be installed and operated in accordance with all of the following:

1. The delay electronics of the delayed egress locking system shall deactivate upon actuation of the *automatic sprinkler system* or automatic fire detection system, allowing immediate, free egress.
2. The delay electronics of the delayed egress locking system shall deactivate upon loss of power controlling the lock or lock mechanism, allowing immediate free egress.
3. The delayed egress locking system shall have the capability of being deactivated at the fire command center and other approved locations.
4. An attempt to egress shall initiate an irreversible process that shall allow such egress in not more than 15 seconds when a physical effort to exit is applied to the egress side door hardware for not more than 3 seconds. Initiation of the irreversible process shall activate an audible signal in the vicinity of the door. Once the delay electronics have been deactivated, rearming the delay electronics shall be by manual means only.

   Exception: Where *approved*, a delay of not more than 30 seconds is permitted on a delayed egress door.
5. The egress path from any point shall not pass through more than one delayed egress locking system.

   Exceptions:
   1. In Group I-2 or I-3 occupancies, the egress path from any point in the building shall pass through not more than two delayed egress locking systems provided the combined delay does not exceed 30 seconds.
   2. In Group I-1 or I-4 occupancies, the egress path from any point in the building shall pass through not more than two delayed egress locking systems provided the combined delay does not exceed 30 seconds and the building is equipped throughout with an automatic sprinkler system in accordance with Section 903.3.1.1.
6. A sign shall be provided on the door and shall be located above and within 12 inches (305 mm) of the door exit hardware:

   6.1. For doors that swing in the direction of egress, the sign shall read: PUSH UNTIL ALARM SOUNDS. DOOR CAN BE OPENED IN 15 [30] SECONDS.

   6.2. For doors that swing in the opposite direction of egress, the sign shall read: PULL UNTIL ALARM SOUNDS. DOOR CAN BE OPENED IN 15 [30] SECONDS.

   6.3. The sign shall comply with the visual character requirements in ICC A117.1.

   Exception: Where *approved*, in Group I occupancies, the installation of a sign is not required where care recipients who, because of clinical needs, require restraint or containment as part of the function of the treatment area.

7. Emergency lighting shall be provided on the egress side of the door.
8. The delayed egress locking system units shall be *listed* in accordance with UL 294.

CHANGE SIGNIFICANCE: Several changes occurred in these revisions. First, rather than prohibiting delayed egress in specific occupancies, the occupancies are listed where delayed egress is allowed. Within this list are two new locations where delayed egress can be installed.

One of these locations is courtrooms. If the courtroom is classified as a Group B occupancy because the occupant load is below 50, then delayed egress was allowed by the 2015 *International Fire Code*. The new exception to Section 1010.1.9.8 allows for delayed egress to be installed within a courtroom classified as Group A-3. This exception only allows delayed egress on the exit doors or exit access doors which are not the main exit. This typically will be the doors on the judge's side of the bar. Note that for use in the Group A courtrooms, a fire sprinkler system is required throughout the building. This is different than other locations for delayed egress where a fire sprinkler system or fire detection system is required. The fire detection system cannot be used in the Group A courtroom as a condition to allow delayed egress.

Delayed egress is also allowed in Group E classrooms, but only when the occupant load is less than 50. Delayed egress is often used to prevent wandering or elopement of younger children.

The new exception in Section 1010.1.9.8.1, Item 5 allows the egress path in Group I-1 and I-4 occupancies to pass through two delayed egress systems. As in Group I-2 occupancies, many Group I-1 occupancies need more than one delayed egress system. For example, if the Group I-1 occupancy is on the 2nd floor, or higher, in a building, a delayed egress system may be needed on the door to the exit stairway on that floor, with another delayed egress locking system at the door to the exterior on the ground floor. In Groups I-1 and I-4, an additional delayed egress locking system may be desired to help reduce wandering or elopement by occupants.

1010.1.9.9, 1010.1.9.10

Electrically Locked Egress Doors

International Code Council®

This pair of doors is electrically locked and provided with sensor release hardware.

CHANGE TYPE: Clarification

CHANGE SUMMARY: Criteria for electrically locked egress doors have been clarified and correlated.

2018 CODE: **1010.1.9.9 Sensor release of electrically locked egress doors.** ~~The electric locks on sensor released doors located in a means of egress in buildings with an occupancy in Group A, B, E, I-1, I-2, I-4, M, R-1 or R-2 and entrance doors to tenant spaces in occupancies in Group A, B, E, I-1, I-2, I-4, M, R-1 or R-2 are permitted~~ <u>Sensor release of electric locking systems shall be permitted on doors located in a *means of egress* in any occupancy except Group H</u> where installed and operated in accordance with all of the following criteria:

1. The sensor shall be installed on the egress side, arranged to detect an occupant approaching the doors <u>and shall cause the electric locking system to unlock.</u>

<u>2.</u> The ~~doors~~ <u>electric locks</u> shall be arranged to unlock by a signal from or loss of power to the sensor.

~~2.~~<u>3.</u> Loss of power to the lock or locking system shall automatically unlock the ~~doors~~ <u>electric locks.</u>

~~3.~~<u>4.</u> The doors shall be arranged to unlock from a manual unlocking device located 40 inches to 48 inches (1016 mm to 1219 mm) vertically above the floor and within 5 feet (1524 mm) of the secured doors. Ready access shall be provided to the manual unlocking device and the device shall be clearly identified by a sign that reads "PUSH TO EXIT." When operated, the manual unlocking device shall result in direct interruption of power to the <u>electric</u> lock—independent of other electronics—and the ~~doors~~ <u>electric lock</u> shall remain unlocked for not less than 30 seconds.

~~4.~~<u>5.</u> Activation of the building fire alarm system, where provided, shall automatically unlock the ~~doors~~ <u>electric lock</u>, and the ~~doors~~ <u>electric lock</u> shall remain unlocked until the fire alarm system has been reset.

~~5.~~<u>6.</u> Activation of the building *automatic sprinkler system* or fire detection system, where provided, shall automatically unlock the ~~doors~~ <u>electric lock</u>. The ~~doors~~ <u>electric lock</u> shall remain unlocked until the fire alarm system has been reset.

~~6.~~<u>7.</u> The door locking system units shall be listed in accordance with UL 294.

1010.1.9.10 ~~Electromagnetically~~ <u>Door hardware release of electrically</u> locked egress doors. ~~Doors in the means of egress in buildings with an occupancy in Group A, B, E, I-1, I-2, I-4, M, R-1 or R-2 and doors to tenant spaces in Group A, B, E, I-1, I-2, I-4, M, R-1 or R-2 shall be permitted to be locked with an electromagnetic locking system where equipped with hardware that incorporates a built-in switch and~~ <u>Door</u>

hardware release of electric locking systems shall be permitted on doors in the *means of egress* in any occupancy except Group H where installed and operated in accordance with all of the following:

1. The door hardware that is affixed to the door leaf has an obvious method of operation that is readily operated under all lighting conditions.
2. The door hardware is capable of being operated with one hand and shall comply with Section 1010.1.9.6.
3. Operation of the door hardware directly interrupts the power to the ~~electromagnetic~~ electric lock and unlocks the door immediately.
4. Loss of power to the electric locking system automatically unlocks the door.
5. Where *panic* or *fire exit hardware* is required by Section 1010.1.10, operation of the *panic* or *fire exit hardware* also releases the ~~electromagnetic~~ electric lock.
6. The locking system units shall be *listed* in accordance with UL 294.

CHANGE SIGNIFICANCE: Section 1010.1.9.9 has been revised to exclude Group H occupancies rather than the long string of occupancies where sensor release hardware is allowed. Sensor release of electrically locked doors is designed to provide immediate egress by sensing the approaching occupant and releasing the electric lock on the door. In many applications, the occupant is unaware that the door is electrically locked as the electrical locks unlock prior to the occupant reaching the door.

The previous language specifying that sensor release locks could only be installed on entrance doors to the building or entrance doors to tenant spaces has been removed. Sensor release electric locks are allowed on any door within the exit path. Terminology was also revised to specify what triggers the release of the electric lock and what actions are necessary for resetting the lock.

Item 1 in Section 1010.1.9.9 is revised to describe the action required, when the sensor detects an approaching occupant, of unlocking the door. There are times when the door needs to unlock, but not unlatch the door. This situation occurs on fire-protection-rated doors.

Section 1010.1.9.10 is revised to clarify the section and correlate with other code sections addressing types of locking systems. The primary difference between Section 1010.1.9.9 "sensor release of electrically locked egress doors" and Section 1010.1.9.10 "door hardware release of electrically locked egress doors" is the initial method of electric lock release. "Sensor release" detects an approaching occupant and releases the door, while "door hardware release" unlocks the door upon manual action of the door mounted hardware such as panic hardware, fire exit hardware, or lever action handles.

1010.1.9.9, 1010.1.9.10 continues

1010.1.9.9, 1010.1.9.10 continued

The requirement in the charging section which required that the door hardware incorporate a built-in switch has been deleted. This was not necessary when one considers that Item 6 requires that the locking system is listed.

The reference to electromagnetic locks is removed from both sections. While electromagnetic locking systems may be the most common, they are not the only type of electric locking hardware. These sections now address all types of electrical locking systems.

1010.1.9.12

Locks on Stairway Doors

CHANGE TYPE: Modification

CHANGE SUMMARY: The limitation is removed which prohibited locking doors on the stairway side when the stairway was more than four stories, but less than a high-rise.

2018 CODE: **1010.1.9.12 Stairway doors.** Interior *stairway means of egress* doors shall be openable from both sides without the use of a key or special knowledge or effort.

Exceptions:

1. *Stairway* discharge doors shall be openable from the egress side and shall only be locked from the opposite side.
2. This section shall not apply to doors arranged in accordance with Section 403.5.3 of the *International Building Code.*
3. ~~In stairways serving not more than four stories,~~ *Stairway* exit doors are permitted to be locked from the side opposite the egress side, provided that they are openable from the egress side and capable of being unlocked simultaneously without unlatching upon a signal from the *fire command center*, if present, or a signal by emergency personnel from a single location inside the main entrance to the building.
4. *Stairway* exit doors shall be openable from the egress side and shall only be locked from the opposite side in Group B, F, M and S occupancies where the only interior access to the tenant space is from a single *exit stairway* where permitted in Section 1006.3.3.
5. *Stairway* exit doors shall be openable from the egress side and shall only be locked from the opposite side in Group R-2

1010.1.9.12 continues

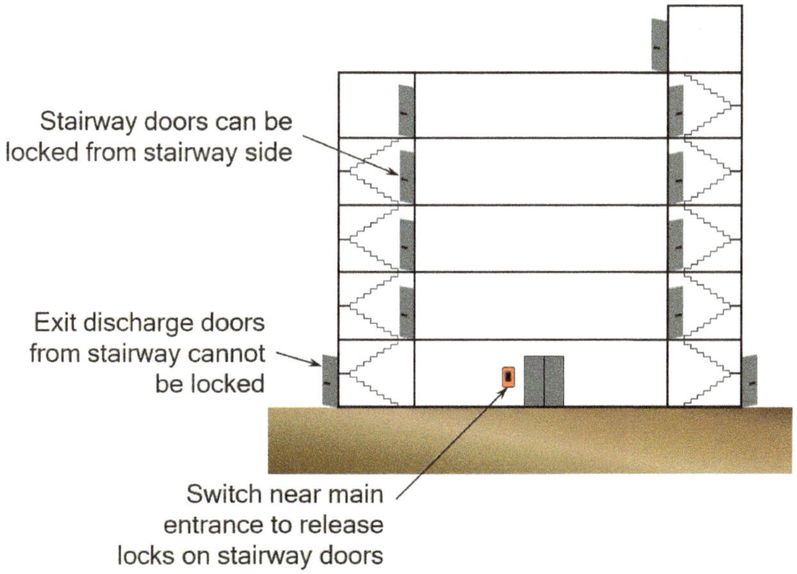

Stairway doors can be locked from the stairway side on the condition that a switch is provided near the main entrance. For high-rise buildings, the switch shall be located in the fire command center.

1010.1.9.12 continued

occupancies where the only interior access to the *dwelling unit* is from a single exit stairway where permitted in Section 1006.3.3.

CHANGE SIGNIFICANCE: The 2015 IFC allows stairway doors to be locked from the side opposite egress on all stories in a high-rise building in Section 403.5.3. For buildings other than high-rise, the 2015 IBC and IFC Section 1010.1.9.11 allowed locking of stairway doors, but the doors in the stairway could only be locked where the stairway served no more than four stories.

Actual application of the code would allow locking stairways up to four stories in height, but at five stories the stairway doors could not be locked. When the building reached high-rise classification, the stairway doors could again be locked.

This revision corrects that anomaly in the code and allows all stairway doors to be locked whether the building is a high-rise or not. For high-rise buildings, a switch must be provided in the fire command center which will unlock, but not unlatch, the stairway doors from the stairway side. For other than high-rise, where there is no fire command center, a switch must still be provided for operation by emergency personnel, but it is now located near the main entrance inside the building.

1010.1.10

Panic Hardware and Fire Exit Hardware

CHANGE TYPE: Modification

CHANGE SUMMARY: Sensor release of electrically locked doors is now allowed on egress doors in Groups A and E. Also, the section is clarified to state that panic hardware or fire exit hardware is only required on swinging doors.

2018 CODE: 1010.1.10 Panic and fire exit hardware. ~~Doors~~ Swinging doors serving a Group H occupancy and swinging doors serving rooms or spaces with an occupant load of 50 or more in a Group A or E occupancy shall not be provided with a latch or lock other than *panic hardware* or *fire exit hardware*.

Exceptions:

1. A main *exit* of a Group A occupancy shall be permitted to be locked in accordance with Section 1010.1.9.4, Item 2.
2. Doors provided with *panic hardware* or *fire exit hardware* and serving a Group A or E occupancy shall be permitted to be ~~electromagnetically~~ electrically locked in accordance with Section 1010.1.9.9 or 1010.1.9.10.

Electrical rooms with equipment rated 1,200 amperes or more and over 6 feet (1829 mm) wide, and that contain overcurrent devices, switching devices or control devices with exit or exit access doors, shall be equipped with *panic hardware* or *fire exit hardware*. The doors shall swing in the direction of egress travel.

International Code Council®

The panic hardware on these exit doors will release the electric lock.

CHANGE SIGNIFICANCE: UL 305 "Panic Hardware" is the standard for testing and listing panic hardware and fire exit hardware. UL 305 applies to outward-opening doors and as such does not apply to the special doors addressed in Section 1010.1.4. However, some have interpreted the current text in Section 1010.1.10 to require panic hardware or fire exit hardware on special doors, such as special purpose horizontal sliding, accordion or folding doors. The revision in the charging statement clarifies that panic hardware and fire exit hardware is required for pivoted or side-hinged swinging doors.

The revision in Exception 2 allows doors with panic or fire exit hardware in Group A or E occupancies to be equipped with hardware complying with Section 1010.1.9.9 for sensor release of electrically locked egress doors, or Section 1010.1.9.10 for door hardware of electrically locked egress doors.

Section 1010.1.9.9 for sensor release egress doors, requires a sensor to detect an occupant approaching the door and cause the electrical locking system to unlock the door, allowing unhindered egress.

Section 1010.1.9.10 for door hardware release egress doors requires the panic or fire exit hardware to cause the electrical locking system to unlock the door, allowing unhindered egress.

1010.1.10 continues

1010.1.10 continued

Both of these locking arrangements facilitate immediate egress either by sensing the approaching occupant or immediately unlocking the electric lock on the door. In many applications, the occupant is unaware that the door is electrically locked.

Both of these locking arrangements release the electric lock when power is lost. For additional information, see Significant Change to Sections 1010.1.9.9 and 1010.1.9.10.

1010.3

Turnstiles

CHANGE TYPE: Addition

CHANGE SUMMARY: This new section allows security turnstiles, or similar barriers, in the means of egress path.

2018 CODE: **1010.3 Turnstiles and similar devices.** Turnstiles or similar devices that restrict travel to one direction shall not be placed so as to obstruct any required *means of egress*, except where permitted in accordance with Sections 1010.3.1, 1010.3.2 and 1010.3.3.

> ~~**Exception:**~~ **1010.3.1 Capacity.** Each turnstile or similar device shall be credited with a capacity based on not more than a 50-person *occupant* load where all of the following provisions are met:
>
> 1. Each device shall turn free in the direction of egress travel when primary power is lost and on the manual release by an employee in the area.
> 2. Such devices are not given credit for more than 50 percent of the required egress capacity or width.
> 3. Each device is not more than 39 inches (991 mm) high.
> 4. Each device has not less than 16½ inches (419 mm) clear width at and below a height of 39 inches (991 mm) and not less than 22 inches (559 mm) clear width at heights above 39 inches (991 mm).

1010.3 continues

International Code Council®

These security barriers are located in the egress path. The egress width is based on the clear width provided when the barriers are in the open position, but can be included as egress width if the proper safety features are provided.

1010.3 continued

1010.3.1.1 Clear width. Where located as part of an *accessible route*, turnstiles shall have not less than 36 inches (914 mm) clear at and below a height of 34 inches (864 mm), not less than 32 inches (813 mm) clear width between 34 inches (864 mm) and 80 inches (2032 mm) and shall consist of a mechanism other than a revolving device.

1010.3.2 Security access turnstiles. Security access turnstiles that inhibit travel in the direction of egress utilizing a physical barrier shall be permitted to be considered as a component of the means of egress, provided that all of the following criteria are met:

1. The building is protected throughout by an approved, supervised automatic sprinkler system in accordance with Section 903.3.1.1.
2. Each security access turnstile lane configuration has a minimum clear passage width of 22 inches (559 mm).
3. Any security access turnstile lane configuration providing a clear passage width of less than 32 inches (810 mm) shall be credited with a maximum egress capacity of 50 persons.
4. Any security access turnstile lane configuration providing a clear passage width of 32 inches (810 mm) or more shall be credited with a maximum egress capacity as calculated in accordance with Section 1005.
5. Each secured physical barrier shall automatically retract or swing to an unobstructed open position in the direction of egress, under each of the following conditions:
   - 5.1. Upon loss of power to the turnstile or any part of the access control system that secures the physical barrier.
   - 5.2. Upon actuation of a clearly identified manual release device with ready access that results in direct interruption of power to each secured physical barrier, after which such barriers remain in the open position for not less than 30 seconds. The manual release device shall be positioned at one of the following locations:
     - 5.2.1. On the egress side of each security access turnstile lane.
     - 5.2.2. At an approved location where it can be actuated by an employee assigned to the area at all times that the building is occupied.
   - 5.3. Upon actuation of the building fire alarm system, if provided, after which the physical barrier remains in the open position until the fire alarm system is manually reset.

     Exception: Actuation of a manual fire alarm box.
   - 5.4. Upon actuation of the building automatic sprinkler or fire detection system, after which the physical barrier remains in the open position until the fire alarm system is manually reset.

~~1010.3.1~~ 1010.3.3 High turnstile. Turnstiles more than 39 inches (991 mm) high shall meet the requirements for revolving doors or the requirements of Section 1010.3.2 for security access turnstiles.

~~1010.3.2~~ 1010.3.4 Additional door. Where serving an *occupant load* greater than 300, each turnstile that is not portable shall have a side-hinged swinging door that conforms to Section 1010.1 within 50 feet (15 240 mm).

> **Exception:** A side-hinged swinging door is not required at security access turnstiles that comply with Section 1010.3.2.

CHANGE SIGNIFICANCE: Manufacturers of turnstile devices have expanded into the security access control market and currently have products that have physical barrier leaves that restrict access into and out of buildings. These devices can vary in height and sophistication to address building security concerns that may not meet safety requirements related to the means of egress. Typically, these turnstile devices are located at building entrances and elevator lobbies. The current requirements for turnstiles apply historically to the "three arm" waist-high turnstiles for entertainment or transportation venues and do not apply to the newer installations. These new provisions provide the code official with criteria to evaluate these new modern turnstiles.

Turnstiles on the market can be as narrow as 22 inches. To be considered for installation in the path of egress, the turnstile or barrier must provide a clear width of at least 22 inches. Barriers between 22 inches and 32 inches can only be considered to accommodate a total of 50 persons. Where barriers provide more than 32 inches, the additional egress width can be included in the total egress width provided.

Turnstiles and barriers which inhibit travel in the direction of egress are allowed if the building is protected with a sprinkler system designed to NFPA 13, "Standard for the Installation of Sprinkler Systems" and the barriers are capable of being retracted or opened automatically or manually by a security guard. Automatic operation would occur upon actuation of the fire sprinkler system required in Item 1, or a fire alarm system, if one is provided.

The manual override for the turnstile access is similar to those requirements for delayed egress locks.

1011.6

Stairway Landings

CHANGE TYPE: Clarification

CHANGE SUMMARY: The method of determining the required width and depth of a stairway landing is clarified.

2018 CODE: **1011.6 Stairway landings.** There shall be a floor or landing at the top and bottom of each *stairway*. The width of landings, measured perpendicularly to the direction of travel, shall be not less than the width of *stairways* served. Every landing shall have a minimum ~~width~~ depth, measured ~~perpendicular~~ parallel to the direction of travel, equal to the width of the *stairway*~~. Where the stairway has a straight run the depth shall be a minimum of~~ or 48 inches (1219 mm), whichever is less. Doors opening onto a landing shall not reduce the landing to less than one-half the required width. When fully open, the door shall not project more than 7 inches (178 mm) into a landing. Where *wheelchair* spaces are required on the *stairway* landing in accordance with Section 1009.6.3, the *wheelchair* space shall not be located in the required width of the landing and doors shall not swing over the *wheelchair* spaces.

> **Exception:** Where *stairways* connect stepped *aisles* to cross *aisles* or concourses, *stairway* landings are not required at the transition between *stairways* and stepped *aisles* constructed in accordance with Section 1029.

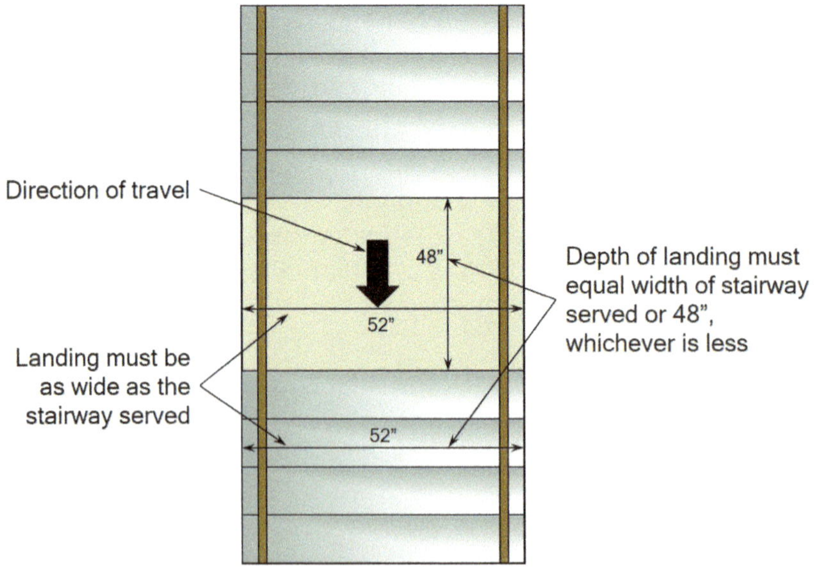

The width of the landing must be at least as wide as the stairway served. The depth must be at least as wide as the stairway served or 48 inches, whichever is less.

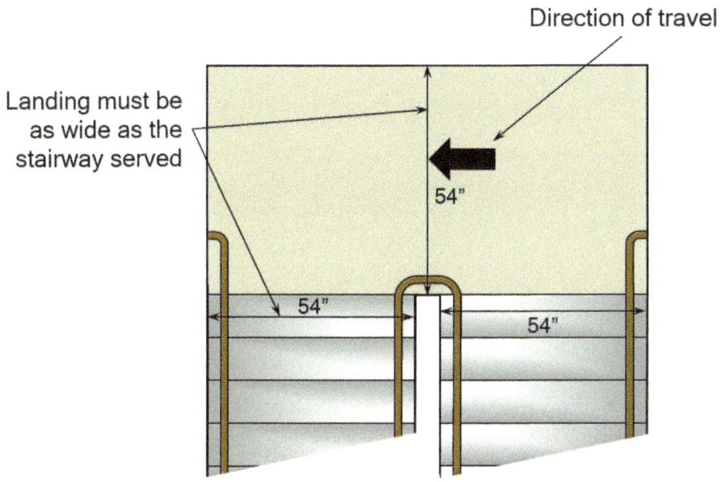

The width of the landing is measured perpendicular to the direction of travel.

CHANGE SIGNIFICANCE: This revised language specifies the minimum depth and width for all stairway landings.

The minimum width of landings is clarified by stating that this is the dimension perpendicular to travel. Similarly, the minimum depth of landings is clarified by referring to it as the dimension parallel to travel.

The width of every landing must be at least the width of the stairway is serves. The minimum depth of a landing must be the lesser of the width of the stairway served or 48 inches.

This modification specifies that the minimum depth of the landing in the direction of travel shall be equal to the width of the stair or 48 inches, whichever is less.

1013.2

Floor-level Exit Signs in Group R-1

International Code Council®

The bottom of the low-level exit sign can be a maximum of 18 inches above the floor.

CHANGE TYPE: Modification

CHANGE SUMMARY: The location of low-level exit signs can now be 18 inches above the floor.

2018 CODE: **1013.2 Floor-level exit signs in Group R-1.** Where exit signs are required in Group R-1 occupancies by Section 1013.1, additional low-level exit signs shall be provided in all areas serving guest rooms in Group R-1 occupancies and shall comply with Section 1013.5.

The bottom of the sign shall be not less than 10 inches (254 mm) nor more than ~~12 inches (305 mm)~~ 18 inches (455 mm) above the floor level. The sign shall be flush mounted to the door or wall. Where mounted on the wall, the edge of the sign shall be within 4 inches (102 mm) of the door frame on the latch side.

CHANGE SIGNIFICANCE: The 2015 IFC provides just a 2-inch tolerance for where the bottom of required low-level exit signs must be located. This 2-inch window is often challenging for designers and property owners due to field conditions or desired interior finish and trim. For example, several high-end resort properties have installed 12-inch tall base boards in the exit access corridors of the hotels. The requirement for the bottom of the sign to be located within 10 to 12 inches above the floor level creates issues for these facilities.

The proposed amendment is to allow the bottom of the required low-level exit signs to be located between 10 and 18 inches above the floor level. The additional 6 inches provides sufficient "wiggle room" for designers and owners. Further, there is no impact on the level of life safety of the occupants of the Group R-1 occupancies since the low-level exit signs will still be visible below the smoke layer from a fire (in the zone in which the occupants would presumably be crawling.)

NFPA 101, *Life Safety Code*® Section 7.10.1.6 permits the bottom of low-level exit signs to be installed between 6 and 18 inches above the floor level. Therefore, there is another code that allows the bottom of the low-level exit signs to be installed up to 18 inches above the floor level. Although NFPA 101 permits the bottom of the low-level exit signs to be just 6 inches above the floor level, the IFC retains the requirement for 10 inches above the floor level because ICC A117.1, Section 404.2.9 requires door surfaces within 10 inches of the floor to be a smooth surface for the full width of the door. The low-level exit signs can be installed either on the wall or the door.

1015.6, 1015.7

Fall Arrest for Rooftop Equipment

CHANGE TYPE: Modification

CHANGE SUMMARY: The specific criteria in the code on fall arrest systems is removed and the ANSI/ASSE Z395.1 standard now governs the installation.

2018 CODE: **1015.6 Mechanical equipment, systems and devices.** *Guards* shall be provided where various components that require service are located within 10 feet (3048 mm) of a roof edge or open side of a walking surface and such edge or open side is located more than 30 inches (762 mm) above the floor, roof or grade below. The *guard* shall extend not less than 30 inches (762 mm) beyond each end of such components. The *guard* shall be constructed so as to prevent the passage of a sphere 21 inches (533 mm) in diameter.

> **Exception:** *Guards* are not required where personal fall arrest anchorage connector devices that comply with ANSI/ASSE Z 359.1 are ~~affixed for use during the entire roof covering lifetime. The devices shall be reevaluated for possible replacement when the entire roof covering is replaced. The devices shall be placed not more than 10 feet (3048 mm) on center along hip and ridge lines and placed not less than 10 feet (3048 mm) from the roof edge or open side of the walking surface~~ installed.

1015.7 Roof access. *Guards* shall be provided where the roof hatch opening is located within 10 feet (3048 mm) of a roof edge or open side of a walking surface and such edge or open side is located more than 30 inches (762 mm) above the floor, roof or grade below. The *guard* shall be constructed so as to prevent the passage of a sphere 21 inches (533 mm) in diameter.

> **Exception:** Guards are not required where ~~permanent~~ personal fall arrest anchorage connector devices that comply with ANSI/ASSE Z 359.1

1015.6, 1015.7 continues

International Code Council®

Anchor systems for fall arrest are specified in ANSI/ASSE Z395.1.

1015.6, 1015.7 continued

are ~~affixed for use during the entire roof covering lifetime. The devices shall be reevaluated for possible replacement when the entire roof covering is replaced. The devices shall be placed not more than 10 feet (3048 mm) on center along hip and ridge lines and placed not less than 10 feet (3048 mm) from the roof edge or open side of the walking surface~~ installed.

CHANGE SIGNIFICANCE: These sections in the IFC are intended to integrate with requirements in the *International Mechanical Code*. Section 306.5.1 of the IMC requires work platforms with guards for equipment and appliances installed on roofs with a slope 3 in 12 and greater, thus, the exception to IMC Section 304.11 appears to apply only to roofs that are flat and up to 2 in 12 slope.

IMC Section 304.11 is revised also as follows:

304.11 Guards. Guards shall be provided where various components that require service and roof hatch openings are located within 10 feet (3048 mm) of a roof edge or open side of a walking surface and such edge or open side is located more than 30 inches (762 mm) above the floor, roof or grade below. The guard shall extend not less than 30 inches (762 mm) beyond each end of components that require service. The top of the guard shall be located not less than 42 inces (1067 mm) above the elevated surface adjacent to the guard. The guard shall be constructed so as to prevent the passage of 21-inch-diameter (533 mm) sphere and shall comply with the loading requirements for guards specified in the *International Building Code*.

Exception: Guards are not required where ~~permanent~~ fall arrest/restraint anchorage connector devices that comply with ANSI/ASSE Z 359.1 are ~~affixed for use during the entire roof covering lifetime. The devices shall be reevaluated for possible replacement when the entire roof covering is replaced. The devices shall be placed not more than 10 feet (3048 mm) on center along hip and ridge lines and placed not less than 10 feet (3048 mm) from the roof edge or open side of the walking surface~~ installed.

By removing the language requiring placement of anchors along hip or ridge lines and along roof edges, all three sections now simply refer to ANSI/ASSE Z359.1 "Safety Requirements for Personal Fall Arrest Systems, Subsystems and Components, Part of the Fall Protection Code."

IFC Section 102.7.1 specifies that when a conflict occurs between the code and a referenced standard, the code language applies. Previously, the anchors had to be installed at intervals of ten feet based on the IFC criteria, even if the standard did not require the same installation. Removing the criteria in the code allows the standard to govern for all installations.

This eliminates any confusion and allows the anchors for each building to be designed based on the actual roof system and equipment location. The standard provides guidance on anchor placement and installation. This allows the standard to be applied on a case by case basis to fit the specific activities that may occur on the individual roof.

1017.3

Common Path of Egress Travel

CHANGE TYPE: Clarification

CHANGE SUMMARY: Common path of egress travel must be applied to each room or space on every story.

2018 CODE: **1017.3 Measurement.** *Exit access* travel distance shall be measured from the most remote point ~~within a story~~ of each room, area or space along the natural and unobstructed path of horizontal and vertical egress travel to the entrance to an *exit*.

> **Exception:** In open parking garages, *exit access* travel distance is permitted to be measured to the closest riser of an *exit access stairway* or the closest slope of an *exit access ramp*.

1017.3 continues

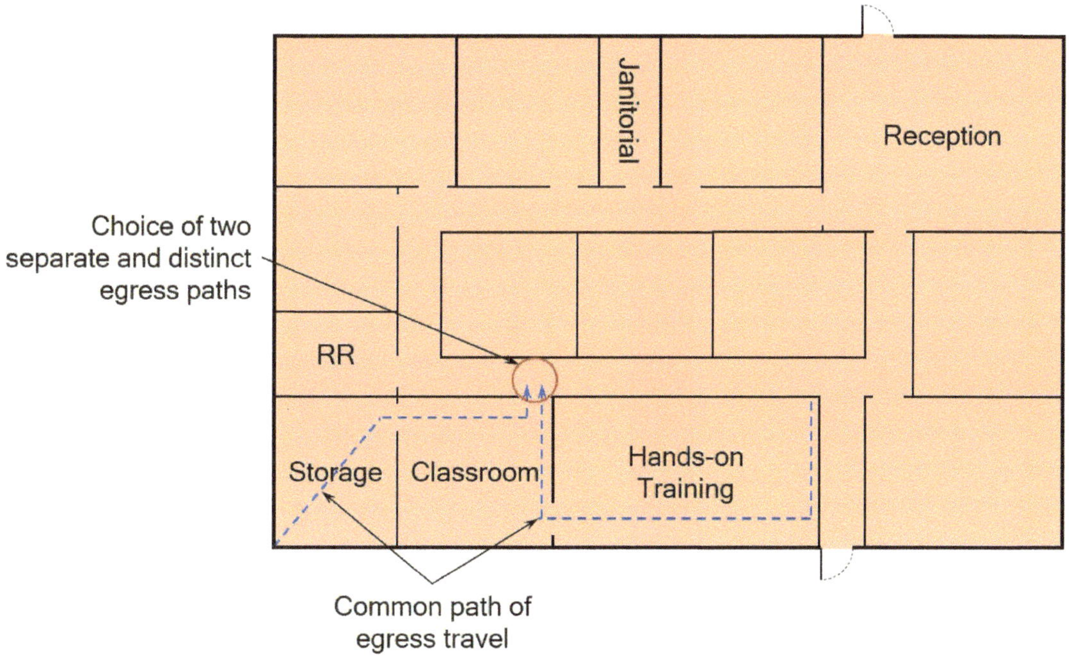

The common path of egress is applied to each room with only one exit or exit access door.

1017.3 continued

SECTION 202 DEFINITIONS

COMMON PATH OF EGRESS TRAVEL. That portion of ~~the~~ *exit access* travel distance measured from the most remote point ~~within a story~~ of each room, area or space to that point where the occupants have separate and distinct access to two *exits* or *exit access* doorways.

CHANGE SIGNIFICANCE: This section is revised along with the definition of "common path of egress travel" to clarify that each room or space must be evaluated. If applied literally, the 2015 IFC could be interpreted such that the common path of egress travel need only be considered from one point (the most remote) on a given story. The common path of travel must be considered in each room of a story, and the exit access travel distance must be considered for each story. The exit access travel distance is measured from the most remote point of a story, but there are designs where the common path of travel must be evaluated in a room adjacent to the interior exit stairway.

Additionally, the deletion of the reference to within a story allows for the common path of travel to continue to an adjacent level, as provided in IFC Section 1006.3. Common path of travel requirements could potentially apply to a multilevel design condition.

Both common path of travel and exit access travel distance are measured in the same manner. In a room or space with only one exit access doorway, the terminus of each is different, but the starting point and the route are the same. The common path of travel is included in the allowed exit access travel distance.

The definition is further refined by specifying that common path of egress travel is measured to the point where separate and distinct egress paths are available. The reference to "separate and distinct" access to exits or exit access doorways means that the egress paths must be independent.

All potential paths of egress travel must be considered when considering common path of egress travel and exit access travel distance. These revisions are designed to clarify that the path of travel originating from any room, area, or space should be evaluated when determining common paths of egress travel.

1023.3.1

Stairway Extension

CHANGE TYPE: Modification

CHANGE SUMMARY: Fire-resistance-rated separation is not required between an interior exit stairway and exit passageways if stairway pressurization is provided.

2018 CODE: **1023.3.1 Extension.** Where *interior exit stairways* and *ramps* are extended to an *exit discharge* or a *public way* by an *exit passageway*, the *interior exit stairway* and *ramp* shall be separated from the *exit passageway* by a *fire barrier* constructed in accordance with Section 707 of the *International Building Code* or a *horizontal assembly* constructed in accordance with Section 711 of the *International Building Code*, or both. The *fire-resistance rating* shall be not less than that required for the *interior exit stairway* and *ramp*. A *fire door* assembly complying with Section 716 of the *International Building Code* shall be installed in the *fire barrier* to provide a *means of egress* from the *interior exit stairway* and *ramp* to the *exit passageway*. Openings in the *fire barrier* other than the *fire door* assembly are prohibited. Penetrations of the *fire barrier* are prohibited.

Exceptions:

1. Penetrations of the *fire barrier* in accordance with Section 1023.5 shall be permitted.
2. Separation between an *interior exit stairway* or *ramp* and the *exit passageway* extension shall not be required where there are no openings into the *exit passageway* extension.
3. Separation between an *interior exit stairway* or *ramp* and the *exit passageway* extension shall not be required where the *interior exit stairway* and the *exit passageway* extension are pressurized in accordance with Section 909.20.5 of the *International Building Code.*

1023.3.1 continues

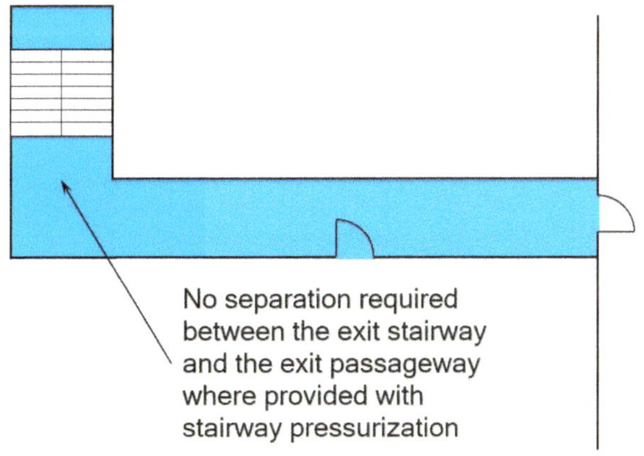

Where stairway pressurization is provided, a fire-resistance-rated separation between an interior exit stairway and exit passageway is not required. Since the rated separation is not required, neither is the wall or door.

1023.3.1 continued

CHANGE SIGNIFICANCE: Interior exit stairways often discharge through an exit passageway to reach the exterior. When the interior exit stairway is pressurized, in accordance with IBC Section 909.20.5, the exit passageway is also typically required to be pressurized since it is a continuation of the pressurized stair enclosure. A single mechanical system is often provided to pressurize the exit stairway and exit passageway. Technical compliance with the code would require separate systems, if a separation is required to be maintained. The introduction of a door and fire barrier between the exit passageway and the stair creates an obstruction to airflow which inhibits the combined pressurization of the stairway and passageway.

The addition of the fire-resistance-rated separation between the stairway and passageway does not provide any added level of safety, and could actually impede egress. With the separation not required, the door and wall are not required. This creates a single atmosphere for the exit stairway and exit passageway.

Note, one of the anomalies of the codes is that while Chapter 9 is very nearly identical in the IFC and the IBC, Section 909.20, addressing smokeproof enclosures, only exists in the IBC. Section 909.20 in the IFC addresses maintenance. When using Exception 3, it is critical that the user refer to Section 909.20 in the *International Building Code*.

1023.5, 1024.6

Exit Stairway and Exit Passageway Penetrations

CHANGE TYPE: Modification

CHANGE SUMMARY: Security system and two-way communication system components are allowed to penetrate the fire-resistant-rated enclosure of exit passageways and interior exit stairways and ramps.

2018 CODE: **1023.5 Penetrations.** Penetrations into or through *interior exit stairways* and *ramps* are prohibited except for the following:

1. Equipment and ductwork necessary for independent ventilation or pressurization.
2. ~~sprinkler piping, standpipes~~ *Fire protection systems.*
3. Security systems.
4. Two-way communication systems.
5. Electrical raceway for fire department communication systems. ~~and~~
6. Electrical raceway serving the interior *exit stairway* and *ramp* and terminating at a steel box not exceeding 16 square inches (0.010 m^2).

Such penetrations shall be protected in accordance with Section 714 of the *International Building Code*. There shall not be penetrations or communication openings, whether protected or not, between adjacent *interior exit stairways* and *ramps*.

Exception: Membrane penetrations shall be permitted on the outside of the *interior exit stairway* and *ramp*. Such penetrations shall be protected in accordance with Section 714.4.2 of the *International Building Code*.

1024.6 Penetrations. Penetrations into or through an *exit passageway* are prohibited except for the following:

1. Equipment and ductwork necessary for independent pressurization.
2. ~~sprinkler piping, standpipes~~ Fire protection systems.
3. Security systems.
4. Two-way communication systems.
5. Electrical raceway for fire department communication.
6. Electrical raceway serving the *exit passageway* and terminating at a steel box not exceeding 16 square inches (0.010 m^2).

Such penetrations shall be protected in accordance with Section 714 of the *International Building Code*. There shall not be penetrations or communicating openings, whether protected or not, between adjacent *exit passageways*.

Exception: Membrane penetrations shall be permitted on the outside of the *exit passageway*. Such penetrations shall be protected in accordance with Section 714.4.2 of the *International Building Code*.

International Code Council®

Components for two-way communication are allowed to penetrate the interior exit stairway enclosure. Note that the membrane penetration for the electrical outlet in the wall is only permitted on the exterior of the enclosure.

1023.5, 1024.6 continues

1023.5, 1024.6 continued

CHANGE SIGNIFICANCE: These two sections are reformatted into a list as the allowed penetrations increase. The specific listing of sprinkler piping and standpipes are included under the general listing of fire protection systems.

Two items added to the list are security systems and two-way communication systems. Building security systems, including cameras in stairways, are becoming more prevalent. If properly protected, a limited number of penetrations for security systems will not result in an unacceptable level of safety. NFPA 101, *Life Safety Code®* requires stairway video monitoring in high-rise buildings with an occupant load of 4,000 or more persons.

Two-way communication systems are required for areas of refuge. While an area of refuge is not required in sprinklered buildings, there are situations where these communication systems may be installed.

The inclusion of these items allows for these systems to be provided in the exit enclosure to provide for safety and security of the building and still maintain the integrity of the exit enclosure.

1025.1

Luminous Egress Path Marking in Group I Occupancies

CHANGE TYPE: Modification

CHANGE SUMMARY: Luminous egress path marking is no longer required in high-rise buildings classified as Groups I-2, I-3, and I-4.

2018 CODE: **1025.1 General.** *Approved* luminous egress path markings delineating the exit path shall be provided in high-rise buildings of Group A, B, E, ~~I~~ I-1, M, and R-1 occupancies in accordance with this section.

> **Exception:** Luminous egress path markings shall not be required on the *level of exit discharge* in lobbies that serve as part of the exit path in accordance with Section 1028.1, Exception 1.

International Code Council®

Luminous egress path marking within an interior exit stairway in a high-rise building.

CHANGE SIGNIFICANCE: The 2015 *International Fire Code* requires luminous egress path marking in all Group I high-rise buildings. This requirement has been revised for Group I high-rise to only require luminous egress path markings in Group I-1 occupancies.

Groups I-2 and I-3 were removed because hospitals, nursing homes, jails, and detention facilities have trained staff that operate with a defend-in-place strategy. The emergency generators are continually monitored and maintained, so the change of the emergency egress lighting required in the means of egress failing is extremely minimal. For the luminous egress path marking to be utilized, both the normal power for means of egress lighting and the emergency generator have to fail.

Group I-4 was also removed. For the requirement for luminous path marking to apply, the high-rise building would need to be classified as a Group I-4. A day care facility is not typically designed as a high-rise building.

1026.4, 1026.4.1

Refuge Areas for Horizontal Exits

CHANGE TYPE: Modification

CHANGE SUMMARY: Guidance is provided to allow enhanced security measures yet still meet egress requirements on classroom doors.

2018 CODE: **1026.4 Refuge area.** The refuge area of a *horizontal exit* shall be a space occupied by the same tenant or a public area and each such refuge area shall be adequate to accommodate the original *occupant load* of the refuge area plus the *occupant load* anticipated from the adjoining compartment. The anticipated *occupant load* from the adjoining compartment shall be based on the capacity of the *horizontal exit* doors entering the refuge area, or the total occupant load of the adjoining compartment, whichever is less.

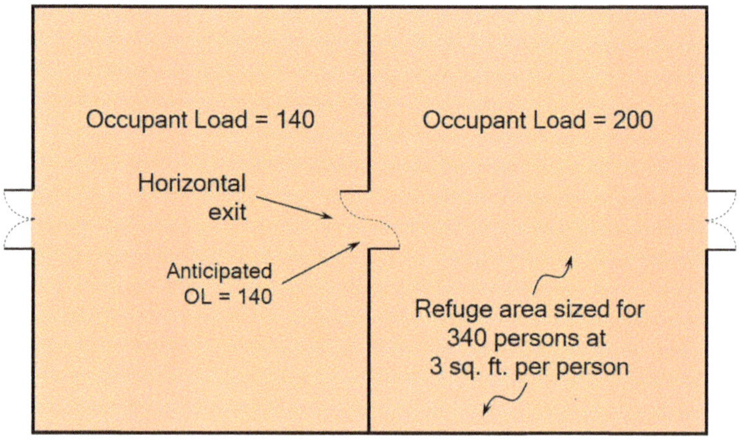

The entire occupant load of the adjacent space is added when determining the size of the refuge area.

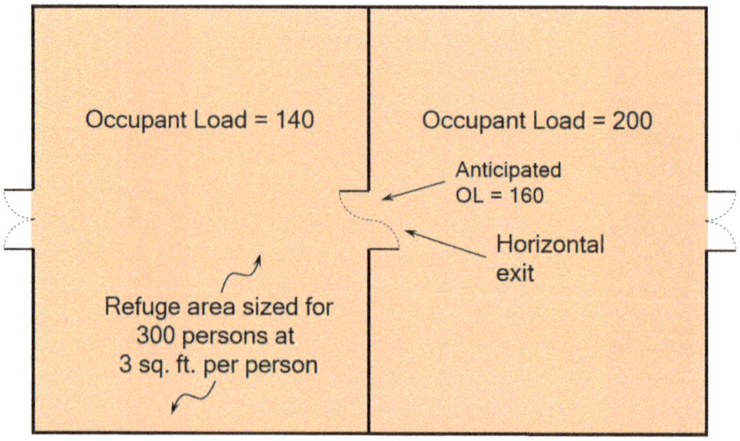

The additional occupant load from the adjacent space is limited by the egress of the horizontal exit when determining the size of the refuge area.

1026.4.1 Capacity. The capacity of the refuge area shall be computed based on a net floor area allowance of 3 square feet (0.2787 m²) for each occupant to be accommodated therein. Where the horizontal exit also forms a smoke compartment, the capacity of the refuge area for Group I-1, I-2 and I-3 occupancies and Group B ambulatory care facilities shall comply with Sections 407.5.3, 408.6.2, 420.6.1 and 422.3.2 of the *International Building Code* as applicable.

> **Exceptions:** ~~The net floor area allowable per occupant shall be as follows for the indicated occupancies:~~
>
> 1. ~~Six square feet (0.6 m²) per occupant for occupancies in Group I-3.~~
> 2. ~~Fifteen square feet (1.4 m²) per occupant for ambulatory occupancies in Group I-2.~~
> 3. ~~Thirty square feet (2.8 m²) per occupant for nonambulatory occupancies in Group I-2.~~

CHANGE SIGNIFICANCE: If a room or space utilizes a horizontal exit as one of the exits, the question arises as to what portion of the occupant load of the original space is included in the cumulative occupant load. Should the occupant load simply be divided by the number of exits or exit access doorways to determine the occupant load contribution to the adjoining space? Based on the distribution of the required capacity within the space under consideration, that may or may not be an appropriate number.

Section 1026.4 has been clarified to indicate that the anticipated occupant load is based on the total occupant load, but not to exceed capacity of the horizontal exit.

A single egress door is required to provide a minimum clear width of 32 inches which, at 0.2 inches per person, can serve 160 occupants. When the occupant load of the space is more than 160, then the horizontal exit becomes the limiting factor. When the occupant load of the space is less than 160, then the actual occupant load is the factor to be used. Note that it will be the entire occupant load, even when the horizontal exit is only one exit of multiple exits available from the space. The entire occupant load will be used, up to the limit of the horizontal exit.

One of the concepts of requiring multiple exits is so that when one egress path is blocked, there is another option. In the situation where the other egress path is blocked, the entire occupant load will be using the horizontal exit. Sizing the refuge area based on that occupant load allows for that situation.

Section 1026.4.1 is also revised by relocating the exceptions into the text and replacing the required dimensions with a reference to the code sections where the dimensions can be found. Since the defend-in-place strategy is now included in the code, these items are no longer exceptions but are accepted and required methods of providing egress in Group I-1, I-2 and I-3 occupancies and Group B Ambulatory Care Facilities. This provides coordination of the requirements, as the requirements change in those sections dealing specifically with the specific occupancy.

1029.6, 1029.6.3, 1029.7

Open-air Assembly Seating

Other Sections affected, but not included herein: 1005.3.1, 1005.3.2, 1009.3, 1009.4, 1019.3, 1029.8, 1029.8.1, 1029.9.5, 1029.12.2.1, Table 1029.12.2.1, 1029.12.2.2

CHANGE TYPE: Modification

CHANGE SUMMARY: A new term and definition is added for open-air assembly seating.

2018 CODE: Due to the extent of the revisions, only specific sections are included here. For the remainder of the text, see the 2018 *International Fire Code.*

1029.6 Capacity of aisle for assembly. The required capacity of *aisles* shall be not less than that determined in accordance with Section 1029.6.1 where *smoke-protected assembly seating* is not provided, and with Section 1029.6.2 ~~or 1029.6.3~~ where *smoke-protected assembly seating* is provided, and with Section 1029.6.3 where open-air assembly seating is provided.

1029.6.3 ~~Outdoor smoke-protected~~ Open-air assembly seating. ~~The~~ In *open-air assembly seating*, the required capacity in inches (mm) of *aisles* shall be not less than the total *occupant load* served by the egress element multiplied by 0.08 (2.0 mm) where egress is by stepped *aisle* and multiplied by 0.06 (1.52 mm) where egress is by level *aisles* and ramped *aisles*.

> **Exception:** The required capacity in inches (mm) of *aisles* shall be permitted to comply with Section 1029.6.2 for the number of seats in the ~~outdoor smoke-protected~~ *open-air assembly seating* where Section 1029.6.2 permits less capacity.

International Code Council®

Open-air assembly seating is addressed in Section 1029.6.

1029.7 Travel distance. Exits and aisles shall be so located that the travel distance to an exit door shall be not greater than 200 feet (60 960 mm) measured along the line of travel in nonsprinklered buildings. Travel distance shall be not more than 250 feet (76 200 mm) in sprinklered buildings. The *exit access* travel distance shall comply with Section 1017. Where *aisles* are provided for seating, the distance shall be measured along the *aisles* and *aisle accessways* without travel over or on the seats.

Exceptions:

1. In facilities with *smoke-protected assembly* seating: The the total *exit access* travel distance shall be not greater than 400 feet (122 m). That portion of the total permitted *exit access* travel distance from each seat to the nearest entrance to a vomitory or concourse shall not exceed 200 feet (60 960 mm). The portion of the total permitted *exit access* travel distance from the entrance to the vomitory or concourse to a stairway, ramp or walk on the exterior of the building one of the following shall not exceed 200 feet (60 960 mm):

   1.1. The closest riser of an *exit access stairway.*

   1.2. Open-air The closest slope of an *exit access ramp.*

   1.3. An exit.

2. In facilities with *open-air assembly seating*: The of Type III, IV or V construction, the total exit access travel distance from each seat to the building exterior one of the following shall not exceed 400 feet (122 m).

   2.1. The closest riser of an *exit access stairway.*

   2.2. The closest slope of an *exit access ramp.*

   2.3. An exit.

3. In facilities with *open-air assembly seating* of Type I or II construction, the *exit access* travel distance shall not be limited. in facilities of Type I or II construction.

SECTION 202 DEFINITIONS

OPEN-AIR ASSEMBLY SEATING. Seating served by *means of egress* that is not subject to smoke accumulation within or under a structure and is open to the atmosphere.

SMOKE-PROTECTED ASSEMBLY SEATING. Seating served by means of egress that is not subject to smoke accumulation within or under a structure for a specified design time by means of passive design or by mechanical ventilation.

CHANGE SIGNIFICANCE: The 2015 IFC uses three terms: "smoke-protected assembly seating," "outdoor smoke-protected assembly seating" and "open-air seating." The revisions to these sections result in the use of only "smoke-protected assembly seating" and "open-air assembly seating."

1029.6, 1029.6.3, 1029.7 continues

1029.6, 1029.6.3, 1029.7 continued

A definition is added for open-air assembly seating, along with clarifying the definition for smoke-protected assembly seating. The main distinction between these definitions is that open-air assembly seating is open to the atmosphere. These two definitions and the coordination of the terms distinguish between the two types of systems that provide smoke protection for assembly seating.

A number of sections were revised to simply use the correct term without any change in intent or application of the code.

Section 1029.7 is revised to clarify its application. The 200- and 250-feet exit access travel distances, for nonsprinklered building and sprinklered buildings, are already in Table 1017.2, so a simple reference to Section 1017 covers these situations. The method of measuring the path of travel is addressed in Section 1017.3; therefore, it has been deleted.

The exceptions are reworded for consistency and correct code terminology. In addition, there is a concern over consistent interpretation. The text of the current exception contains terminology that could be interpreted as always measuring travel distance to the building exit at grade. Historically, where smoke-protected seating or open-air assembly seating is provided, these facilities have permitted the use of open stairways where the means of egress is open to the outside.

Finally, Exception 3 is not new; it was extracted from Exception 2. Placing this exception into a standalone exception ensures that it will not be missed. Exception 3 allows for an unlimited exit access travel distance in buildings of Type I or II construction.

1029.9.1

Minimum Aisle Width

CHANGE TYPE: Modification

CHANGE SUMMARY: Minimum aisle widths in assembly occupancies are clarified with a reference added for minimum widths for accessible routes.

2018 CODE: **1029.9.1 Minimum aisle width.** The minimum clear width for *aisles* shall comply with one of the following:

1. Forty-eight inches (1219 mm) for stepped *aisles* having seating on ~~each side~~ both sides.

   Exception: Thirty-six inches (914 mm) where the stepped *aisles* serve less than 50 seats.

2. Thirty-six inches (914 mm) for stepped *aisles* having seating on only one side.

   Exception: Twenty-three inches (584 mm) between a stepped aisle *handrail* and seating where a stepped *aisle* does not serve more than five rows on one side.

3. Twenty-three inches (584 mm) between a stepped aisle *handrail* or *guard* and seating where the stepped aisle is subdivided by a mid-aisle *handrail.*

4. Forty-two inches (1067 mm) for level or ramped *aisles* having seating on both sides.

   Exceptions:
   1. Thirty-six inches (914 mm) where the *aisle* serves less than 50 seats.
   2. Thirty inches (762 mm) where the *aisle* serves fewer than 15 seats and does not serve ~~more than 14 seats~~ as part of an accessible route.

5. Thirty-six inches (914 mm) for level or ramped *aisles* having seating on only one side.

   Exception: ~~For other than ramped aisles that serve as part of an accessible route, 30~~ Thirty inches (762 mm) where the ~~ramped~~

1029.9.1 continues

1029.9.1 continued

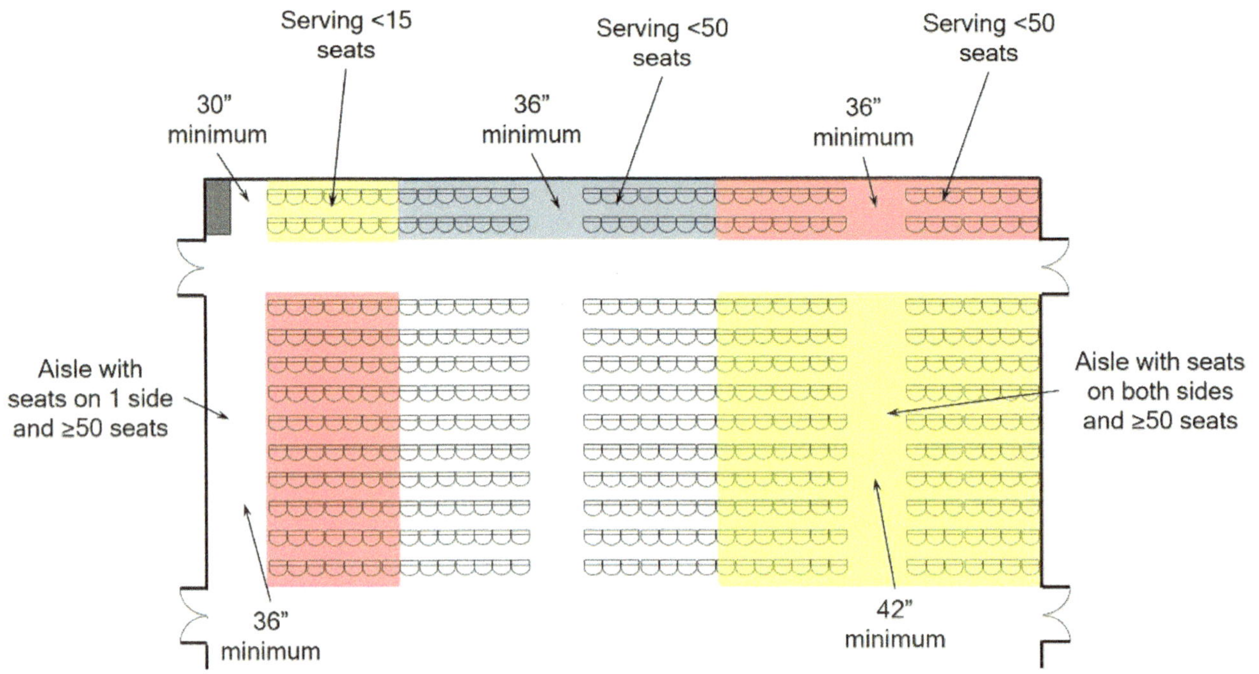

The aisle width can be reduced to 30 inches where the aisle is not required to be an accessible aisle and it serves less than 15 seats.

aisle serves fewer than 15 seats and does not serve ~~more than 14 seats~~ as part of an *accessible route.*

CHANGE SIGNIFICANCE: The provisions have been revised to clarify the section and provide consistency with the code language.

Item 1 is revised with no technical change. It is revised to match the language in Item 4.

Exception 2 of Item 4 contains two revisions. First, it is revised by changing the language "does not serve more than 14 seats" to "serves fewer than 15 seats." This is not a technical change, but it provides consistency with the language in the other exceptions. Second, a reminder regarding accessible routes is added. Sections 1009 and 1104 require that an accessible route with a minimum width of 36 inches be provided to the wheelchair spaces for ingress and egress. Therefore, Item 4, Exception 2 allows the width to reduce to 30 inches, but only if the route is not part of an accessible route.

The exception to Item 5 has been revised so it is similar to the language in Item 4. Additionally, the word "ramped" is deleted. Item 5 addresses both ramped aisles and level aisles. However, the exception was previously limited to "ramped" aisles. Since level aisles are less hazardous than ramped aisles, the exception should also apply to level aisles. This has been corrected by deleting the word "ramped."

1030.1

Emergency Escape and Rescue Openings

CHANGE TYPE: Modification

CHANGE SUMMARY: Emergency escape and rescue openings are required in Groups R-3 and R-4, and Group R-2 provided with only one means of egress from a story. Also, it is possible to eliminate some, or all, emergency escape and rescue openings from a sprinklered basement.

2018 CODE: **1030.1 General.** In addition to the *means of egress* required by this chapter, ~~provisions shall be made for~~ *emergency escape* and *rescue openings* shall be provided in the following occupancies:

1. Group R-2 occupancies located in ~~accordance~~ stories with only one exit or access to only one exit as permitted by Tables ~~1006.3.2(1)~~ 1006.3.3(1) and ~~1006.3.2(2)~~ 1006.3.3(2). ~~and~~
2. Group R-3 and R-4 occupancies.

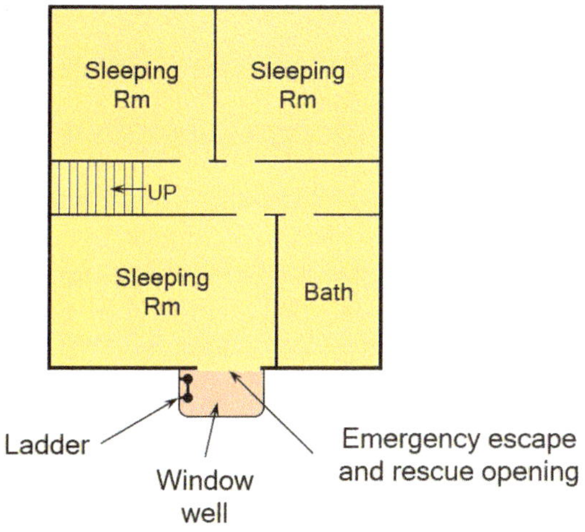

Only one emergency escape and rescue opening is required when the basement of this Group R-3 occupancy is equipped with an automatic sprinkler system.

Basements and sleeping rooms below the fourth story above *grade plane* shall have not fewer than one exterior *emergency escape and rescue opening* in accordance with this section. Where *basements* contain one or more sleeping rooms, *emergency escape and rescue openings* shall be required in each sleeping room, but shall not be required in adjoining areas of the *basement*. Such openings shall open directly into a *public way* or to a *yard* or *court* that opens to a *public way*.

Exceptions:

1. *Basements* with a ceiling height of less than 80 inches (2032 mm) shall not be required to have *emergency escape and rescue openings*.
2. *Emergency escape and rescue openings* are not required from *basements* or sleeping rooms that have an *exit door* or *exit*

1030.1 continues

1030.1 continued

access door that opens directly into a *public way* or to a *yard*, *court* or exterior exit balcony that opens to a *public way*.

3. *Basements* without *habitable spaces* and having not more than 200 square feet (18.6 m^2) in floor area shall not be required to have *emergency escape and rescue openings*.
4. Within individual dwelling and sleeping units in Groups R-2 and R-3, where the buildings is equipped throughout with an automatic sprinkler system installed in accordance with Section 903.3.1.1, 903.3.1.2 or 903.3.1.3, sleeping rooms in *basements* shall not be required to have emergency escape and rescue openings provided that the *basement* has one of the following:
    - 4.1. One means of egress and one emergency escape and rescue opening.
    - 4.2. Two means of egress.

CHANGE SIGNIFICANCE: Several revisions occurred in the charging language in Section 1030.1.

- The section is revised by creating a list separating the requirements for Group R-2 and Group R-3.
- The provision for Group R-2 is revised by specifying that it only applies if the occupancy is located on a story with only one exit. This requirement is in the 2015 IBC, but it is only found when the tables are consulted and the reader realizes that they only apply to occupancies with one exit. The inclusion of that language in this section no longer forces the reader to go back to Section 1006.3.2 to determine applicability, so interpreting and applying this section is more consistent.
- Group R-4 is added in the list. Group R-4 is required to follow design and construction requirements for Group R-3, so adding it here is only for clarification.

Exception 4 is added to address Groups R-2 and R-3 buildings equipped with an automatic sprinkler system. This exception states that basements in sprinklered buildings of Groups R-2 or R-3 are allowed to provide either two means of egress, or one means of egress and one emergency escape and rescue opening. Already, the code allows Groups R-1 and Group I occupancies to eliminate emergency escape and rescue openings in sprinklered basements. This is not an equivalent provision to those situations, but it allows the elimination of an emergency escape and rescue opening in each bedroom in a basement of Groups R-2 and R-3. This exception requires at least one basement escape window or door, or an additional means of egress in addition to the primary means of egress. This exception only applies if the building is equipped with an automatic sprinkler system throughout.

There were three individual code changes that provided the revisions to this section. One of the code changes added Group R-4 to the charging paragraph. Group R-4 was not included in Exception 4 because it was added in a separate code change. IBC Section 310.5 states that a Group R-4 is to be constructed as required for Group R-3. Therefore, it would be appropriate to include Group R-4 in application of the new Exception 4.

1030.1.1

Operation of Emergency Escape and Rescue Openings

CHANGE TYPE: Addition

CHANGE SUMMARY: Fall prevention devices are allowed on emergency escape and rescue openings provided that they comply with ASTM F2090.

2018 CODE: **1030.1.1 Operational constraints and opening control devices.** Emergency escape and rescue openings shall be operational from inside the room without the use of keys or tools. Window-opening control devices complying with ASTM F 2090 shall be permitted for use on windows serving as a required emergency escape and rescue opening.

~~**1030.5**~~ **1030.4 Window wells.** *No change in text...*

~~**1030.5.1**~~ **1030.4.1 Minimum size.** *No change in text...*

~~**1030.5.2**~~ **1030.4.2 Ladders or steps.** *No change in text...*

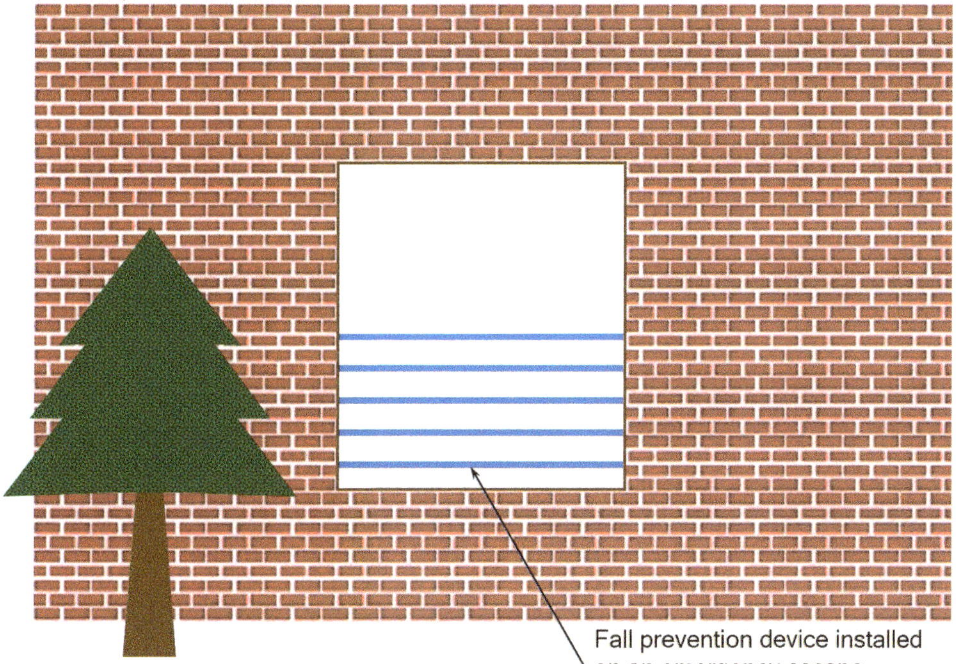

The fall prevention device must be openable since the window is an emergency escape and rescue opening.

~~**1030.4**~~ **1030.5** ~~**Operational constraints**~~ **Bars, grilles, covers and screens.** ~~Emergency escape and rescue openings shall be operational from the inside of the room without the use of keys or tools.~~ Bars, grilles, ~~grates~~ covers, screens or similar devices are permitted to be placed over *emergency escape and rescue openings*, bulkhead enclosures, or window wells that serve such openings, provided that the minimum net clear opening size complies with ~~Section 1030.2~~ Sections 1030.1.1 through 1030.4.2 and such devices shall be releasable or removable from the inside without

1030.1.1 continues

1030.1.1 continued

the use of a key, tool or force greater than that which is required for normal operation of the escape and rescue opening. Where such bars, grilles, ~~grates~~ covers, screens or similar devices are installed in existing buildings, ~~smoke alarms shall be installed in accordance with Section 907.2.10 regardless of the valuation of the alteration.~~ they shall not reduce the net clear opening of the *emergency escape and rescue opening* and *smoke alarms* shall be installed in accordance with Section 907.2.10 regardless of the valuation of the *alteration*.

CHANGE SIGNIFICANCE: Section R310.1.1 of the 2015 *International Residential Code*® states that devices complying with ASTM F2090 "Specification for Window Fall Prevention Devices with Emergency Escape (Egress) Release Mechanisms" do not jeopardize compliance with the emergency escape and rescue provisions. A similar requirement has now been added to the IFC in Section 1030.1.1.

Section 1030.1.1 deals with fall protection devices installed over a window which is also an emergency escape and rescue opening. Therefore, the device must be stable enough to prevent falls, and easily removable or openable to allow for escape. ASTM F2090 allows no more than two actions to open the fall prevention device to allow for escape.

Section 1030.5 has been relocated to the end of this series of requirements to allow for a reference back to Sections 1030.1.1 through 1030.4.2 regarding the design and location of the opening. The first sentence is also deleted since it is stated again halfway through the section.

Section 1030.5 has also been revised to include a reference to bulkhead enclosures. The minimum clear opening size, in Section 1030.2, of 5.7 square feet can be provided with either a door or a window. If a window is installed, a window well is provided to provide access to ground level. If a door is installed, a bulkhead enclosure is needed to provide access to ground level.

1031.2.2

Locking Arrangements in Existing Educational Occupancies

CHANGE TYPE: Addition

CHANGE SUMMARY: Guidance is provided to allow enhanced security measures yet still meet egress requirements on classroom doors.

2018 CODE: **1031.2.2 Locking arrangements in educational occupancies.** In Group E occupancies, Group B educational occupancies and Group I-4 occupancies, egress doors from classrooms, offices and other occupied rooms shall be permitted to be provided with locking arrangements designed to keep intruders from entering the room where all of the following conditions are met:

1. The door shall be capable of being unlocked from outside the room with a key or other approved means.
2. The door shall be openable from within the room in accordance with Section 1010.1.9.
3. Modifications shall not be made to existing listed panic hardware, fire door hardware or door closers.
4. Modifications to fire door assemblies shall be in accordance with NFPA 80.

Photo courtesy of ASSA ABLOY

A fob is used to operate this classroom security lock from inside the room.

CHANGE SIGNIFICANCE: A high priority in many locations is the safety of occupants at schools and in classrooms during the event of a threatening situation. These revisions provide criteria in the code which balance the challenges of providing protection for students and teachers in the classroom and at the same time provide for free and immediate egress.

These locks are added into existing buildings to provide security, but they must also allow free egress and comply with the single operation requirement. These provisions provide guidance for combining security while maintaining safe egress.

Many of the traditional locksets required the instructor to first open the classroom door to lock the door with a key from the hallway side, then reenter the classroom for a defend-in-place strategy. This action places the instructor at risk by forcing them to open the door and become exposed in the hallway. The classroom security function can be activated from the interior, eliminating the need to open the classroom door while still allowing unrestricted egress from inside the classroom.

Section 1031.2.2 is added to the code and provides guidance for the installation of locks in existing Group E and B educational occupancies and Group I-4 day care facilities.

Item 1 requires that the door shall be unlockable from outside the classroom. This allows for fire fighters, law enforcement, or school personnel to obtain entry even when the door is locked from the inside. Several options are currently available to provide for unlocking from the exterior, such as key, card reader, pass code, or even remotely operated from an approved location. For new construction, Section 1010.1.4.4.1 allows remote operation of classroom locks. This means that the doors can be locked electrically from a single location. This design allows all of the

1031.2.2 continues

1031.2.2 continued

classroom doors to be locked, and unlocked, with the operation of a single switch or fob. This switch can be designed to be key operated to add an additional level of security. Even though that section is not referenced here, it could be considered as another "approved means."

Item 2 indicates that the lock mechanism must comply with Section 1010.1.9, which includes Section 1010.1.9.5 requiring single operation. Therefore, even when the door is locked to secure the door from intruders, it must be openable from the inside with a single action.

Items 3 and 4 ensure listed fire doors and door hardware are maintained so the integrity of the opening protective is maintained.

See also Significant Change to Section 1010.1.4.4 for locking arrangements in new educational occupancies.

1031.4

Exit Signs in Existing Buildings

CHANGE TYPE: Modification

CHANGE SUMMARY: The application of exit sign requirements in existing buildings has been clarified for both installation and maintenance.

2018 CODE: 1031.4 Exit signs. Exit signs shall be installed and maintained in accordance with the building code that was in effect at the time of construction and the applicable provisions in Section ~~1013~~ 1104. Decorations, furnishings, equipment or adjacent signage that impairs the visibility of exit signs, creates confusion or prevents identification of the *exit* shall not be allowed.

CHANGE SIGNIFICANCE: The previous language had been interpreted to require existing buildings to install exit signs to meet the provisions of Section 1013 in the 2015 Fire Code. This meant that every time the requirements were modified in Section 1013, all buildings would need to comply with those revised requirements. Since Section 1031 covers maintenance, the original intent was to require that the exit signs were maintained.

To correct this language, the revision clarifies that existing exit signs must be maintained in accordance with the original code that required their installation and any additional requirements found in Section 1104.

Retroactive construction requirements are contained in Chapter 11. This reference to Section 1104 sends the code user to Sections 1104.3 and 1104.4. Section 1104.3 requires that exits signs have internal or external illumination, or are the self-luminous type of exit sign. Section 1104.4 states that when illumination of the means of egress is required in Section 1104.5, then the exit signs shall be visible under the emergency power illumination. Section 1104.5 also has an exception for self-luminous type exit signs.

The end result is that existing buildings do not need to comply with the current requirements for exit signs, but the exit signs must be illuminated in both normal and emergency power modes.

International Code Council®

Existing exit which must be illuminated under normal lighting and emergency lighting.

1031.10

Inspection and Testing of Emergency Egress Lighting

International Code Council®

Emergency egress lighting is manually tested monthly unless equipped with self-testing and self-diagnostics.

CHANGE TYPE: Modification

CHANGE SUMMARY: Inspection and testing requirements for emergency egress lighting are relocated into Chapter 10 Means of Egress and revised to allow self-diagnostics.

2018 CODE: **~~604.6~~ 1031.10 Emergency lighting equipment inspection and testing.** Emergency lighting shall be maintained in accordance with Section 108 and shall be inspected and tested in accordance with Sections ~~604.6.1 through 604.6.2.1~~ 1031.10.1 and 1031.10.2.

~~604.6.1~~ 1031.10.1 Activation test. ~~An activation~~ Emergency lighting equipment shall be tested monthly for a duration of not less than 30 seconds. The test shall be performed manually or by an automated self-testing and self-diagnostic routine. Where testing is performed by self-testing and self-diagnostics, a visual inspection of the emergency lighting equipment shall be ~~completed~~ conducted monthly to identify any equipment displaying a trouble indicator or that has become damaged or otherwise impaired. ~~The activation test shall ensure the emergency lighting activates automatically upon normal electrical disconnect and stays sufficiently illuminated for not less than 30 seconds.~~

~~**604.6.1.1 Activation test record.** Records of tests shall be maintained. The record shall include the location of the emergency lighting tested, whether the unit passed or failed, the date of the test and the person completing the test.~~

~~**604.6.2.1 Power test record.** Records of tests shall be maintained. The record shall include the location of the emergency lighting tested, whether the unit passed or failed, the date of the test and the person completing the test.~~

~~604.6.1~~ 1031.10.2 Power test. ~~For battery-powered emergency lighting, a power test of the~~ Battery-powered emergency lighting equipment shall be ~~completed~~ tested annually~~. The~~ by operating the equipment on battery power ~~test shall operate the emergency lighting~~ for not less than 90 minutes. ~~and shall remain sufficiently illuminated for the duration of the test.~~

CHANGE SIGNIFICANCE: This change moves requirements for emergency lighting equipment inspection and testing from the general emergency and standby power requirements of Chapter 6 to the means of egress requirements in Chapter 10 where it more logically is found.

Additional changes recognize modern technology found in many emergency lighting units, including self-testing and self-diagnostics to replace manual testing.

Conditions for the annual requirement of a 90-minute duration test have been clarified so the test can occur without having to disconnect normal premises power.

Finally, by referring users to IFC Section 108 Maintenance, recordkeeping requirements are made consistent with the remainder of the fire code.

1103.5.1

Fire Sprinklers in Existing Group A-2 Occupancies

CHANGE TYPE: Addition

CHANGE SUMMARY: A section has been added to Chapter 11 which requires the retrofit installation of a fire sprinkler system in existing Group A-2 occupancies where alcoholic beverages are consumed if the occupant load is 300 or more.

2018 CODE: **1103.5.1 Group A-2.** Where alcoholic beverages are consumed in a Group A-2 occupancy having an occupant load of 300 or more, the fire area containing the Group A-2 occupancy shall be equipped with an *automatic sprinkler system* in accordance with Section 903.3.1.1.

CHANGE SIGNIFICANCE: This new section requires the installation of an automatic sprinkler system in existing Group A-2 occupancies where both of the following conditions occur:

1. The occupant load is 300 or more.
2. Alcoholic beverages are consumed in the facility.

This requirement is located in Chapter 11 since it is a construction requirement applicable to existing buildings. This requirement applies even though the existing building complied with the code it was constructed under.

1103.5.1 continues

International Code Council®

Existing Group A-2 occupancies will be required to install an automatic sprinkler system if the occupant load is 300 or more.

1103.5.1 continued

This code change is a reaction to The Station Nightclub Fire of 2003. The National Institute of Standards and Technology thoroughly investigated the fire and fatalities that resulted. The NIST report (NIST NCSTAR 2: Vol. I) contained seven recommendations. Recommendation #2 of the NIST report states:

> *"NIST recommends that model codes require sprinkler systems, and that state and local authorities adopt and aggressively enforce this provision:*
>
> *a) for all new nightclubs regardless of size, and*
>
> *b) for existing nightclubs with an occupancy limit greater than 100 people."*

Group A-2 occupancies involve conditions such as large occupant loads, high occupant density, significant fuel loading and moveable furnishings and decorations. Group A-2 occupancies also include the potential for reduced lighting levels, high noise levels, combustible decorations, strobe and flashing lights, and confusing egress paths. Each of these alone can be a significant issue, but when combined with the consumption of alcohol they lead to the inability of the occupants to promptly and safely exit the building under fire conditions.

The Station Nightclub Fire is not the only fire to provide justification for this requirement. Unfortunately, fires and fatalities in nightclubs has been a repeat occurrence. The table below lists nightclub fires in the United States where multiple fatalities have occurred.

NAME	LOCATION	DATE	NUMBER OF FATALITIES
Study Club	Detroit, Michigan	1929	22
Rhythm Club	Natchez, Mississippi	1940	209
Cocoanut Grove	Boston, Massachusetts	1942	492
Dale's Penthouse	Montgomery, Alabama	1967	26
Upstairs Lounge	New Orleans, Louisiana	1973	32
Gulliver's Nightclub	Port Chester, New York	1974	24
Beverly Hills Supper Club	Southgate, Kentucky	1977	165
Happy Land	The Bronx, New York	1990	87
The Station Nightclub	West Warwick, Rhode Island	2003	100
Ghost Ship Collective[a]	Oakland, California	2016	36

a. The Ghost Ship Collective fire occurred after the code change was submitted and approved. However, it is further evidence that this catastrophe continues to repeat itself.

While the NIST recommendation is to require fire sprinklers in existing facilities where the occupant load exceeds 100, the new IFC section requires fire sprinklers in Group A-2 occupancies where alcoholic drinks are consumed if the occupant load is 300 or more. The section does not require other fire areas adjacent to the Group A-2 to be sprinklered, nor does it require the entire floor to be sprinklered as would be typical for new construction involving a Group A-2 occupancy.

The fact that this provision is triggered by consumption of alcoholic beverages allows a connection to licensing laws that many jurisdictions typically have in place for sale of such beverages. Such licensing laws, where they apply, will provide significant leverage for jurisdictions to be able to effectively enforce the requirement for a fire sprinkler system as a condition of being code compliant and issued a license.

1103.9

Carbon Monoxide Alarms in Existing Buildings

International Code Council®

Combination smoke alarm and carbon monoxide alarm installed on the ceiling.

CHANGE TYPE: Modification

CHANGE SUMMARY: Carbon monoxide alarms are no longer required to be retroactively installed in existing Groups I-1, I-2, I-4, and R based on occupancy classification. The retroactive installation of carbon monoxide alarms is only required in existing sleeping rooms and dwelling units.

2018 CODE: **1103.9 Carbon monoxide alarms.** ~~Existing Group I-1, I-2, I-4 and R occupancies shall be equipped with carbon monoxide alarms installed in accordance with Section 915, except that~~ Carbon monoxide alarms shall be installed in existing dwelling units and sleeping units where those units include any of the conditions identified in Sections 915.1.2 through 915.1.6. The carbon monoxide alarms shall be ~~allowed to be solely battery operated~~ installed in the locations specified in Section 915.2 and the installation shall be in accordance with Section 915.4.

Exceptions:

1. Carbon monoxide alarms are permitted to be solely battery operated where the code that was in effect at the time of construction did not require carbon monoxide detectors to be provided.
2. Carbon monoxide alarms are permitted to be solely battery operated in dwelling units that are not served from a commercial power source.
3. A carbon monoxide detection system in accordance with Section 915.5 shall be an acceptable alternative to carbon monoxide alarms.

CHANGE SIGNIFICANCE: Most carbon monoxide poisoning fatalities in buildings occur in dwelling units with fuel burning appliances or attached garages. Therefore, these revisions change the requirements from application based on occupancy classification to actual use. The reference to Groups R, I-1, I-2 and I-4 occupancies has been replaced with a reference to dwelling units and sleeping units. As a result, this revision limits the retroactive application to only those portions of buildings where these incidents occur.

This code change also allows the use of battery powered units. This offers a relatively low cost solution to meet the requirement and provide protection in existing buildings.

1104.16.2

Wall Openings Adjacent to Fire Escapes

CHANGE TYPE: Modification

CHANGE SUMMARY: Door and window openings within 10 feet of a fire escape must be protected with ¾-hour opening protectives unless the building is sprinklered.

2018 CODE: 1104.16.2 ~~Protection of openings~~ Opening protectives. ~~Openings~~ Doors and windows within 10 feet (3048 mm) of fire escape *stairways* shall be protected ~~by opening protectives having a minimum~~ with ¾-hour ~~*fire-protection rating*~~ opening protectives.

> **Exception:** ~~In~~ Opening protectives shall not be required in buildings equipped throughout with an approved *automatic sprinkler system*~~, opening protection is not required~~.

CHANGE SIGNIFICANCE: This revision provides an alternative to installing opening protectives in existing buildings which are adjacent to fire escapes. The purpose of IFC Chapter 11 is to provide mitigation of a hazard known to occur in existing buildings which complied with the code at time of construction. Chapter 11 does not intend to require compliance with the current code, but requires a method of mitigating the hazard.

Fire escapes are not allowed as a means of egress in new construction. However, fire escapes have been installed on many buildings in an effort to provide a second exit in multistory buildings with only one exit. As we have learned through fire experience, these exterior stairs need to be protected from the fire within the building.

Under the *International Building Code*, new construction would be required to provide ¾-hour opening protectives for all openings with 10 feet of any portion of an exterior exit stairway. In addition, new exterior stairs require that the exterior wall is of 1-hour construction within 10 feet of the stairway. This section does not require compliance with the current requirements, but rather allows an option to the owner. The door and window openings within 10 feet could be protected, or the building could be provided with an automatic sprinkler system. Note, this does not apply to other openings such as vent terminations.

It should be noted that the *International Existing Building Code*® allows this alternative for Level 2 Alterations in Section 805.3.1.2.1 and general fire escape requirements in Section 504.5.

International Code Council®

All of the windows and doors in this wall are required to be protected with ¾-hour assemblies, or the building can be equipped with an automatic sprinkler system.

1105.6.2

Fire-protection-rated Doors in Existing Group I-2

A hold-open device installed at a smoke compartment separation in an existing building.

CHANGE TYPE: Addition

CHANGE SUMMARY: Fire-protection-rated doors in existing Group I-2 occupancies have three options for automatic closing operations.

2018 CODE: **1105.6.2 Group I-2 occupancies.** In Group I-2, where a door serves as an opening protective in a fire barrier, smoke barrier or fire wall and where the door is equipped with a hold-open device, such door shall automatically close upon any of the following conditions:

1. Actuation of smoke detectors initiating the hold-open device.
2. Activation of the fire alarm system within the zone.
3. Activation of an automatic sprinkler system within the zone.

CHANGE SIGNIFICANCE: This revision provides options for existing doors in fire barrier walls, fire walls and smoke barriers. This section applies to the doors that provide opening protection and are automatic closing upon release of a hold-open device. Doors that are self-closing are not affected by this section.

The three options specify that the doors release upon activation of the smoke detector located adjacent to the doors, activation of a fire alarm device within the fire alarm zone within which the doors are located, or the operation of the fire sprinkler system protecting the area within which the doors are located.

Chapter 12
Energy Systems

Photo courtesy of Ioxus, Inc.

Safety features for this capacitor energy storage system are now included in Chapter 12.

CHANGE TYPE: Addition

CHANGE SUMMARY: This new chapter has been added to the IFC to address all configurations of energy systems. This chapter contains the emergency power, standby power, and stationary battery storage system requirements from Chapter 6 of the 2015 IFC along with new requirements for other methods of energy generation and storage.

2018 CODE: Due to the extent of the code text, an abbreviated outline of the chapter is shown below. For the full text, see the 2018 *International Fire Code.*

CHAPTER 12
ENERGY SYSTEMS

1201 GENERAL

1201.1 Scope.

1201.2 Electrical wiring and equipment.

1201.3 Mixed system installation.

1202 DEFINITIONS

1202.1 Definitions.

1203 EMERGENCY AND STANDBY POWER SYSTEMS

1203.1 General.

1203.2 Where required.

1203.3 Critical circuits.

1203.4 Maintenance.

1203.5 Operational inspection and testing.

1203.6 Supervision of maintenance and testing.

1204 SOLAR PHOTOVOLTAIC POWER SYSTEMS

1204.1 General.

1204.2 Access and pathways.

1204.3 Other than Group R-3 buildings.

1204.4 Ground-mounted photovoltaic panel systems.

1204.5 Buildings with rapid shutdown.

1205 STATIONARY FUEL CELL POWER SYSTEMS

1205.1 General.

1205.2 Permits.

1205.3 Equipment.

Chapter 12 continues

Chapter 12 continued

1205.4 Installation.

1205.5 Residential use.

1205.6 Indoor installations.

1205.7 Vehicle impact protection.

1205.8 Outdoor installation.

1205.9 Fuel supply.

1205.10 Manual shutoff.

1205.11 Ventilation and exhaust.

1205.12 Fire suppression.

1205.13 Gas detection systems.

1206 ELECTRICAL ENERGY STORAGE SYSTEMS

1206.1 Scope.
1206.2 Stationary storage battery systems.
1206.3 Capacitor energy storage systems.

CHANGE SIGNIFICANCE: This new chapter contains several sections from the 2015 IFC and several new items that have been added to the code. The chapter intends to address all systems used to generate or store electrical energy. All of these items have been brought together to create the new Chapter 12 for Energy Systems. The table below identifies the requirements in Chapter 12.

2018 IFC Section	2018 IFC Section	2015 IFC Section
Emergency and standby power systems	1203	604
Solar photovoltaic power systems	1204	605.11
Fuel cell power systems	1205	new
Stationary storage battery systems	1206.2	608.1
Capacitor energy storage systems	1206.3	new

In addition to the relocated requirements, several new provisions have been included in Chapter 12; fuel cell power systems being one of these items. Fuel cell power systems are being used in ever increasing numbers to meet facility energy needs. Stationary fuel cell power systems generate power through an electrochemical process that combines hydrogen and oxygen to produce electricity. The hydrogen comes from a direct hydrogen source or from any hydrocarbon fuel such as natural gas, gasoline, diesel, or methanol if the fuel cell power system includes integral reforming. The oxygen comes from air around the fuel cell. Section 1205 provides a comprehensive set of requirements to mitigate potential hazards associated with the installation and use of stationary fuel cell power systems.

Three referenced documents form the basis for the fuel cell requirements:

1. ANSI/CSA FC 1, "Stationary Fuel Cell Power Systems" is used to investigate and list the stationary fuel cells. The construction and performance requirements in the standard address a variety of

hazards, including mechanical, electrical, thermal, malfunction, erroneous human intervention and environmental.

2. NFPA 853, "Standard for the Installation of Stationary Fuel Cell Power Systems" includes requirements for the design, construction, and installation of stationary fuel cell power systems.
3. NFPA 2, *Hydrogen Technologies Code* covers the production, storage, transfer, and use of hydrogen in all occupancies and on all premises.

Section 1205 includes requirements for the design, construction, and installation of stationary fuel cell power systems which are similar to NFPA 853 requirements. Section 1205.3 allows for the installation of listed systems or field-fabricated power systems.

Section 1206.3 is added to address capacitor energy storage systems. The U.S. Department of Energy is working with a wide range of stakeholders to encourage the development of large scale electrical energy storage systems (ESS). ESS are needed because the amount of electricity that can be generated on the electrical grid is relatively fixed over short periods of time, although demand for electricity fluctuates throughout the day. The intent of this technology is to store electrical energy so it can be available to meet demand as needed. This technology is actively being pursued with a number of energy storage technologies, including battery storage systems and electrochemical capacitors, among others.

The upcoming widespread installation of large ESS systems into existing buildings poses significant hazards to occupants and emergency responders because of the nature of the technologies involved and the large amounts of energy being stored. These systems are largely unregulated by the 2015 editions of the IBC and IFC, especially capacitors energy storage systems (CESS) which have the potential to be a significant energy storage source by 2018.

Section 1206.3 establishes basic protection requirements for CESS. These sections include safety measures that are similar to battery storage system requirements. An overview of the requirements is:

1. Construction and operational permits are required for CESS systems exceeding 3 kWh (10.8 megajoules). See Section 1206.3.
2. CESS below the maximum allowable quantity is allowed in incidental use areas and does not need to comply with hazardous material requirements, which is consistent with current battery storage system requirements.
3. Where a single fire area contains CESS systems exceeding 600 kWh (1260 megajoules), it is classified as a Group H occupancy. See Section 1206.3.3.
4. The protection concepts in Sections 1206.3.1 through 1206.3.2.5 are similar to the battery storage system requirements.
5. Sections 1206.3.4 to 1206.3.6 are new and are similar to new requirements for battery storage system requirements.

Revisions have also occurred in the requirements for solar photovoltaic power systems. See Significant Change to Section 1204.5.

Revisions have also occurred in the requirements for stationary storage battery systems. See Significant Change to Section 1206.2.

1204.5

Rapid Shutdown for Solar Photovoltaic Power Systems

CHANGE TYPE: Addition

CHANGE SUMMARY: Rapid shutdown is required on solar photovoltaic systems to reduce the shock hazard to emergency responders.

2018 CODE: **1204.5 Buildings with rapid shutdown.** Buildings with rapid shutdown solar photovoltaic systems shall have permanent labels in accordance with Sections 1204.5.1 through 1204.5.3.

1204.5.1 Rapid shutdown type. The type of solar photovoltaic system rapid shutdown shall be labeled with one of the following:

1. For solar photovoltaic systems that shut down the array and the conductors leaving the array, a label shall be provided. The first two lines of the label shall be uppercase characters with a minimum height of $^{3}/_{8}$ inch (10 mm) in black on a yellow background. The remaining characters shall be uppercase with a minimum height of $^{3}/_{16}$ inch (5 mm) in black on a white background. The label shall be in accordance with Figure 1204.5.1(1) and state the following:

SOLAR PV SYSTEM EQUIPPED WITH RAPID SHUTDOWN.
TURN RAPID SHUTDOWN SWITCH TO THE "OFF" POSITION TO
SHUT DOWN PV SYSTEM AND REDUCE SHOCK HAZARD IN ARRAY.

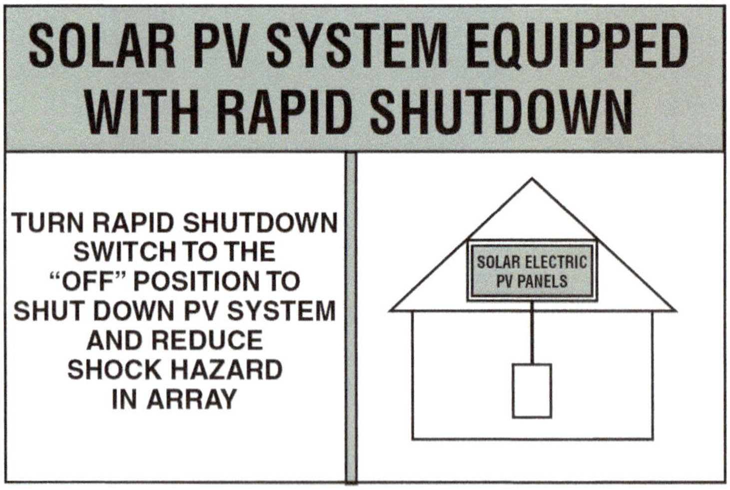

Figure 1204.5.1(1)
LABEL FOR SOLAR PV SYSTEMS THAT REDUCE SHOCK HAZARD WITHIN ARRAY AND SHUT DOWN CONDUCTORS LEAVING ARRAY

2. For ~~solar~~ photovoltaic systems that only shut down conductors leaving the array, a label shall be provided. The first two lines of the label shall be uppercase characters with a minimum height of $^{3}/_{8}$ inch (10 mm) in white on a red background and the remaining

characters shall be capitalized with a minimum height of $^{3}/_{16}$ inch (5 mm) in black on a white background. The label shall be in accordance with Figure 1204.5.1(2) and state the following:

THIS SOLAR PV SYSTEM EQUIPPED WITH RAPID SHUTDOWN.
TURN RAPID SHUT DOWN SWITCH TO THE "OFF" POSITION
TO SHUT DOWN CONDUCTORS OUTSIDE THE ARRAY.
CONDUCTORS WITHIN ARRAY REMAIN ENERGIZED IN SUNLIGHT.

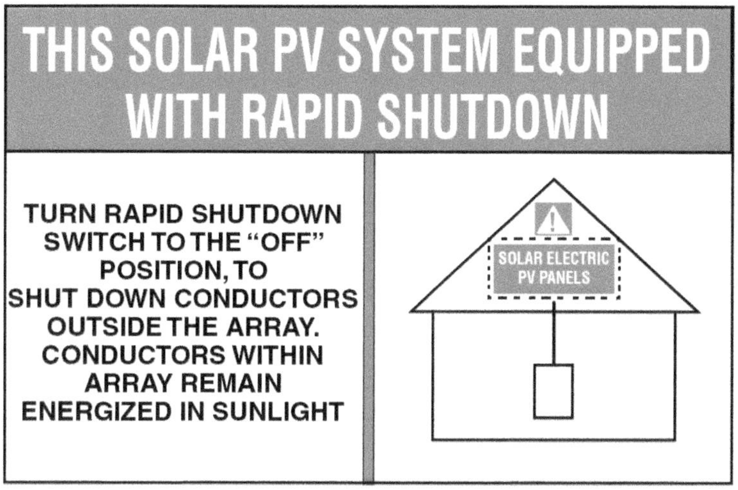

Figure 1204.5.1(2)
LABEL FOR SOLAR PV SYSTEMS THAT ONLY SHUT DOWN CONDUCTORS LEAVING THE ARRAY

1204.5.1.1 Diagram. The labels in Section 1204.5.1 shall include a simple diagram of a building with a roof. Diagram sections in red signify sections of the solar photovoltaic system that are not shut down when the rapid shutdown switch is turned off.

1204.5.1.2 Location. The rapid shutdown label in Section 1204.5.1 shall be located not greater than 3 feet (914 mm) from the service disconnecting means to which the photovoltaic systems are connected, and shall indicate the location of all identified rapid shutdown switches if not at the same location.

1204.5.2 Buildings with more than one rapid shutdown type. Solar photovoltaic systems that contain rapid shutdown in accordance with both Items 1 and 2 of Section 1204.5.1 or solar photovoltaic systems where only portions of the systems on the building contain rapid shutdown, shall provide a detailed plan view diagram of the roof showing each different photovoltaic system and a dotted line around areas that remain energized after the rapid shutdown switch is operated.

1204.5.3 Rapid shutdown switch. A rapid shutdown switch shall have a label located not greater than 3 feet (914 mm) from the switch that states the following:

RAPID SHUTDOWN SWITCH FOR SOLAR PV SYSTEM

1204.5 continues

1204.5 continued

CHANGE SIGNIFICANCE: These new marking requirements for rapid shutdown PV systems were initially developed by the NFPA Fire Fighter Safety and PV Systems Task Group that was reorganized in December of 2014. This collaborative Task Group consists of more than 20 participants from Code Making Panel 4 of the NFPA 70, the solar industry, the fire service, the insurance industry, test laboratories, and other relevant stakeholders.

NFPA 70 *National Electrical Code*, Section 690.12 requires that solar photovoltaic systems installed on or in buildings be provided with a rapid shutdown function. The intent of this device is to reduce the shock hazard for emergency responders. When activated, the rapid shutdown switch must reduce the voltage in the conductors outside of the solar panel array to a maximum of 30 volts within 30 seconds.

Section 1204.5.3 requires a sign that identifies the rapid shutdown switch for the system.

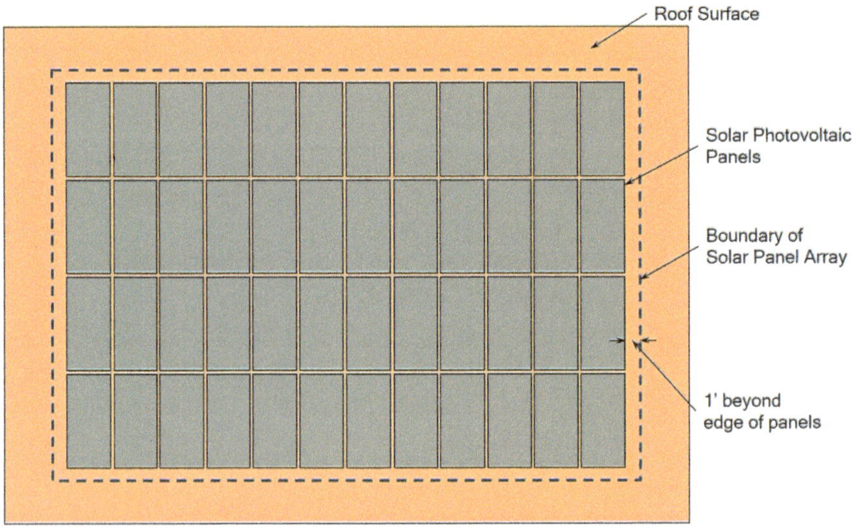

Rapid shutdown must occur for all circuits outside of the boundary of the solar photovoltaic panel array. The boundary is 1 foot outside of the edge of the solar panels.

One of two different signs are required depending on what type of rapid shutdown system is installed on the building. Systems which shut down the conductors leaving the array and the conductors within the array are referenced in Section 1204.5.1, Item 1 and are required to have the sign as shown in Figure 1204.5.1(1). The rapid shut down referenced in Section 1204.5.1, Item 2 only shuts down the conductors leaving the boundary of the solar photovoltaic panel array, so a different sign is required as shown in Figure 1204.5.1(2). NFPA 70 defines the boundary of the solar panel array and it is depicted in the figure above.

CHANGE TYPE: Modification

1206.2

Stationary Storage Battery Systems

CHANGE SUMMARY: This revision moves the stationary battery storage system requirements from Section 608 to Section 1206.2 and includes new battery technologies and required safety features.

2018 CODE: Due to the extent of the code text, only certain sections are shown below. For the full text, see the 2018 *International Fire Code.*

1206 ELECTRICAL ENERGY STORAGE SYTEMS

TABLE 1206.2 Battery Storage System Threshold Quantities

BATTERY TECHNOLOGY	CAPACITY[a]
Flow batteries[b]	20 kWh
Lead acid, all types	70 kWh
Lithium, all types	20 kWh
Nickel cadmium (Ni-Cd)	70 kWh
Sodium, all types	20 kWh[c]
Other battery technologies	10 kWh

For SI:1 kilowatt hour = 3.6 megajoules.
a. For batteries rated in amp-hours, kWh shall equal rated voltage times amp-hour rating divided by 1000.
b. Shall include vanadium, zinc-bromine, polysulfide-bromide and other flowing electrolyte-type technologies.
c. 70 kWh for sodium-ion technologies.

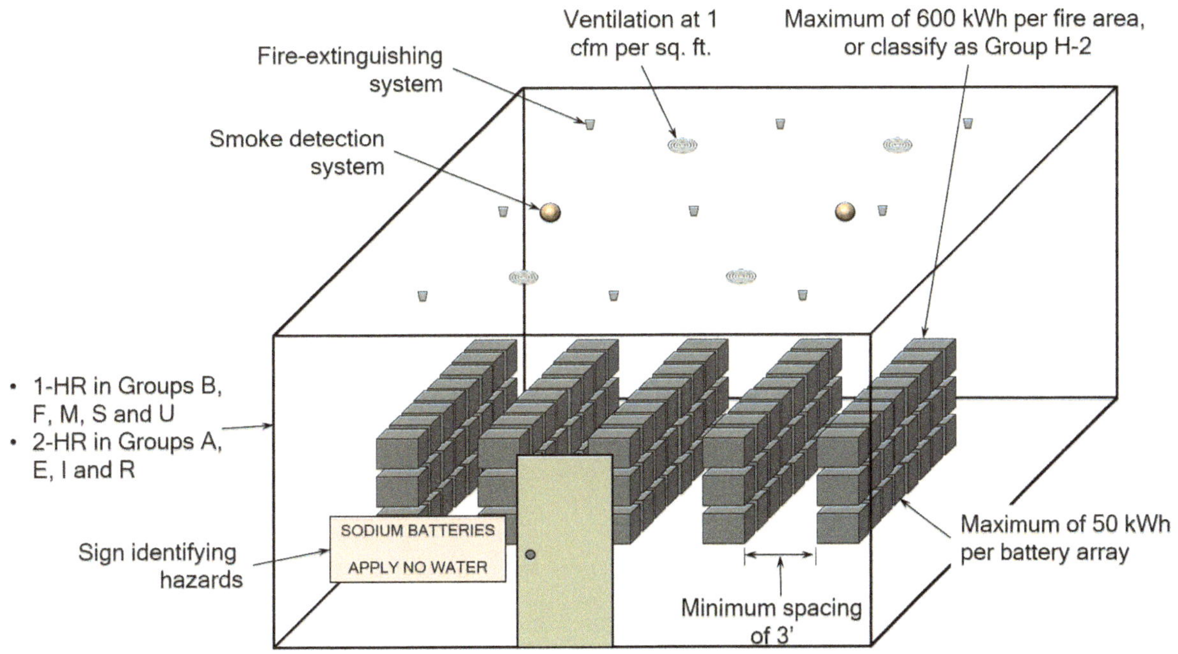

This graphic depicts a stationary storage battery system containing sodium beta batteries.

1206.2 continues

1206.2 continued

1206.2.1 Permits. Permits shall be obtained for the installation and operation of stationary storage battery systems in accordance with Section 105.7.2.

1206.2.3 Hazard mitigation analysis. A failure modes and effects analysis (FMEA) or other approved hazard mitigation analysis shall be provided in accordance with Section 104.7.2 under any of the following conditions:

1. Battery technologies not specifically identified in Table 1206.2 are provided.
2. More than one stationary storage battery technology is provided in a room or indoor area where there is a potential for adverse interaction between technologies.
3. When allowed as a basis for increasing maximum allowable quantities in accordance with Section 1206.2.9.

1206.2.8.3 Stationary battery arrays. Storage batteries, prepackaged stationary storage battery systems and preengineered stationary storage battery systems shall be segregated into stationary battery arrays not exceeding 50 kWh (180 megajoules) each. Each stationary battery array shall be spaced not less than 3 feet (914 mm) from other stationary battery arrays and from walls in the storage room or area. The storage arrangements shall comply with Chapter 10.

Exceptions:

1. Lead acid and nickel cadmium storage battery arrays.
2. Listed preengineered stationary storage battery systems and prepackaged stationary storage battery systems shall not exceed 250 kWh (900 megajoules) each.
3. The fire code official is authorized to approve listed, preengineered and prepackaged battery arrays with larger capacities or smaller battery array spacing if large-scale fire and fault

TABLE 1206.2.9 Maximum Allowable Battery Quantities

BATTERY TECHNOLOGY	MAXIMUM ALLOWABLE QUANTITIES[a]	GROUP H OCCUPANCY
Flow batteries[b]	600 kWh	Group H-2
Lead acid, all types	Unlimited	Not Applicable
Lithium, all types	600 kWh	Group H-2
Nickel cadmium (Ni-Cd)	Unlimited	Not Applicable
Sodium, all types	600 kWh	Group H-2
Other battery technologies	200 kWh	Group H-2[c]

For SI:1 kilowatt hour = 3.6 megajoules.

a. For batteries rated in amp-hours, Kilowatt-hours (kWh) shall equal rated battery voltage times the amp-hour rating divided by 1,000.

b. Shall include vanadium, zinc-bromine, polysulfide-bromide, and other flowing electrolyte-type technologies.

c. Shall be a Group H-4 occupancy if the fire code official determines that a fire or thermal runaway involving the battery technology does not represent a significant fire hazard.

condition testing conducted or witnessed and reported by an approved testing laboratory is provided showing that a fire involving one array will not propagate to an adjacent array, and be contained within the room for a duration equal to the fire-resistance rating of the room separation specified in Table 509 of the *International Building Code.*

1206.2.10.3 Energy management system. An approved energy management system shall be provided for battery technologies other than lead-acid and nickel cadmium for monitoring and balancing cell voltages, currents and temperatures within the manufacturer's specifications. The system shall transmit an alarm signal to an approved location if potentially hazardous temperatures or other conditions such as short circuits, over voltage or under voltage are detected.

1206.2.11.1.1 Alternative fire-extinguishing systems. Battery systems that utilize water-reactive materials shall be protected by an approved alternative automatic fire-extinguishing system in accordance with Section 904. The system shall be listed for protecting the type, arrangement and quantities of storage batteries in the room. The *fire code official* shall be permitted to approve the alternate fire-extinguishing system based on full-scale fire and fault condition testing conducted or witnessed and reported by an *approved* laboratory.

202 DEFINITIONS

~~**BATTERY SYSTEM, STATIONARY LEAD ACID.** A system which consists of three interconnected subsystems:~~

~~1. A lead-acid battery.~~

~~2. A battery charger.~~

~~3. A collection of rectifiers, inverters, converters and associated electrical equipment as required for a particular application.~~

BATTERY SYSTEM, STATIONARY STORAGE. A rechargeable energy storage system consisting of electrochemical storage batteries, battery chargers, controls and associated electrical equipment designed to provide electrical power to a building. The system is typically used to provide standby or emergency power, an uninterruptable power supply, load shedding, load sharing or similar capabilities.

BATTERY TYPES.

Flow battery. A type of storage battery that includes chemical components dissolved in two different liquids. Ion exchange, which provides the flow of electrical current, occurs through the membrane while both liquids circulate in their own respective spaces.

Lead-acid battery. A storage battery that is comprised of lead electrodes immersed in sulfuric acid electrolyte.

Lithium-ion battery. A storage battery ~~that consists of~~ with lithium ions ~~embedded in a carbon graphite or nickel metal-oxide substrate. The electrolyte is a carbonate mixture or a gelled polymer. The lithium ions are the~~ serving as the charge carriers of the battery. The electrolyte is a polymer mixture of carbonates with an inorganic salt and can be in a liquid

1206.2 continues

1206.2 continued

or a gelled polymer form. Lithiated metal oxide is typically a cathode and forms of carbon or graphite typically form the anode.

Lithium metal polymer battery. A storage battery that is comprised of nonaqueous liquid or polymerized electrolytes, which provide ionic conductivity between lithiated positive active material electrically separated from metallic lithium or lithiated negative active material.

Nickel-cadmium (Ni-Cd) battery. An alkaline storage battery in which the positive active material is nickel oxide, the negative contains cadmium and the electrolyte is potassium hydroxide.

Nonrecombinant battery. A storage battery in which, under conditions of normal use, hydrogen and oxygen gasses created by electrolysis are vented into the air outside of the battery.

Preengineered stationary storage battery system. An energy storage system consisting of batteries, a battery management system, components and modules that are produced in a factory, designed to comprise the system when assembled on the job site.

Prepackaged stationary storage battery system. An energy storage system consisting of batteries, a battery management system, components and modules that is factory assembled and shipped as a complete unit for installation at the job site.

Recombinant battery. A storage battery in which, under conditions of normal use, hydrogen and oxygen gases created by electrolysis are converted back into water inside the battery instead of venting into the air outside of the battery.

Sodium-beta storage battery. A storage battery, also referred to as Na-beta batteries or NBB, which uses a solid beta-alumina electrolyte membrane that selectively allows sodium ion transport between a positive electrode such as metal halide and a negative sodium electrode.

Stationary storage battery. A group of electrochemical cells interconnected to supply a nominal voltage of DC power to a suitably connected electrical load, designed for service in a permanent location. The number of cells connected in a series determines the nominal voltage rating of the battery. The size of the cells determines the discharge capacity of the entire battery. After discharge, it may be restored to a fully charged condition by an electric current flowing in a direction opposite to the flow of current when the battery is discharged.

Valve-regulated lead-acid (VRLA) battery. A lead-acid battery consisting of sealed cells furnished with a valve that opens to vent the battery whenever the internal pressure of the battery exceeds the ambient pressure by a set amount. In VRLA batteries, the liquid electrolyte in the cells is immobilized in an absorptive glass mat (AGM cells or batteries) or by the addition of a gelling agent (gel cells or gelled batteries).

Vented (flooded) lead-acid battery. A lead-acid battery consisting of cells that have electrodes immersed in liquid electrolyte. Flooded lead-acid batteries have a provision for the user to add water to the cell and are equipped with a flame-arresting vent which permits the escape of hydrogen and oxygen gas from the cell in a diffused manner such that a spark,

~~or other ignition source, outside the cell will not ignite the gases inside the cell.~~

ENERGY MANAGEMENT SYSTEMS. An electronic system that protects stationary storage batteries from operating outside their safe operating parameters, and generates an alarm and trouble signal for off normal conditions.

STATIONARY BATTERY ARRAY. An arrangement of individual stationary storage batteries in close proximity to each other, mounted on storage racks or in modules, battery cabinets or other enclosures.

CHANGE SIGNIFICANCE: Stationary battery storage systems have been relocated from Section 608 of the 2015 *International Fire Code* to the new Chapter 12 and several revisions have occurred in the regulations. Not all of the revisions are addressed in this text, see the 2018 IFC for the full text. See also Significant Change to Chapter 12.

Several new battery types now fall under these requirements. This includes flow batteries, sodium-beta batteries, and an open category for other technologies not listed. The technology for batteries and energy storage is advancing rapidly and the latter category will allow for future development of these technologies.

Section 1206.2.1 specifies that a construction permit is now required in addition to the operational permit. The construction permit is required to ensure that the safety features and protection measures are incorporated into the design and construction of battery storage systems.

A new Table 1206.1 has been included which changes the method to determine if a battery system must comply with these requirements. Previously, the threshold was either 50 gallons (189 L) for flooded lead-acid, nickel cadmium (Ni-Cd) and valve-regulated lead-acid (VRLA), or 1,000 pounds (454 kg) for lithium-ion and lithium metal polymer. The threshold is now based on the electrical output of the system. This results in the regulation of lead-acid and nickel cadmium (Ni-Cd) systems at 70 kWh and lithium, sodium and flow battery systems at 20 kWh. All other technologies will be regulated when they exceed 10 kWh.

Section 1206.2.3 now requires a report evaluating the failure modes and effects analysis when:

- the battery system falls into the "other battery technology" category, or
- when multiple types of battery systems are located within a room or area, or
- when the applicant desires to increase the quantities above those allowed in Section 1206.2.9.

Section 1206.2.9 is new to the code and it provides a limit on the size of the battery system that can be installed without classification as a Group H occupancy. Table 1206.2.9 establishes the maximum size. The allowable size is based again on electrical output with no limit for lead-acid and nickel cadmium (Ni-Cd) systems. However, when lithium, sodium or flow battery systems exceed 600 kWh, they must be placed into

1206.2 continues

1206.2 continued

a Group H-2 occupancy. All other technologies will be located in a Group H-2 when they exceed 200 kWh.

Even when the limits in Table 1206.2.9 are not exceeded, Section 1206.2.8.3 limits the size of each battery array to a maximum of 50 kWh for sodium and lithium based batteries. Each array is then separated by a minimum of 3 feet to other battery arrays. There are exceptions for listed battery systems and prepackaged battery systems not exceeding 250 kWh.

Battery systems, other than lead-acid and Ni-Cd, must be provided with an energy management system. This system must monitor the voltage, current and temperatures of each battery. When hazardous temperatures or conditions are detected, it must send an alarm to an approved location.

Section 1206.2.11.1 requires an automatic sprinkler system in rooms containing a stationary storage battery system. However, an exception refers back to Section 903.2 which allows spaces or areas in telecommunications buildings used exclusively for telecommunications equipment and associated equipment and batteries to be protected with a smoke detection system and separated from the remainder of the building by 1-hour fire barriers and 2-hour horizontal assemblies in lieu of a sprinkler system. Additionally, Section 1206.2.11.1.1 allows the use of alternate fire-extinguishing systems for protecting batteries that present a water-reactive hazard.

PART 4

Special Occupancies

Chapters 20 through 39

- **Chapter 20** Aviation Facilities No changes addressed
- **Chapter 21** Dry Cleaning No changes addressed
- **Chapter 22** Combustible Dust-Producing Operations
- **Chapter 23** Motor Fuel-Dispensing Facilities and Repair Garages
- **Chapter 24** Flammable Finishes
- **Chapter 25** Fruit and Crop Ripening No changes addressed
- **Chapter 26** Fumigation and Insecticidal Fogging No changes addressed
- **Chapter 27** Semiconductor Fabrication Facilities No changes addressed
- **Chapter 28** Lumber Yards and Agro-industrial, Solid Biomass and Woodworking Facilities
- **Chapter 29** Manufacture of Organic Coatings No changes addressed
- **Chapter 30** Industrial Ovens No changes addressed
- **Chapter 31** Tents, Temporary Special Event Structures and Other Membrane Structures
- **Chapter 32** High-piled Combustible Storage
- **Chapter 33** Fire Safety During Construction and Demolition
- **Chapter 34** Tire Rebuilding and Tire Storage No changes addressed
- **Chapter 35** Welding and Other Hot Work No changes addressed
- **Chapter 36** Marinas No changes addressed
- **Chapter 37** Combustible Fibers
- **Chapter 38** Higher Education Laboratories
- **Chapter 39** Processing and Extraction Facilities
- **Chapters 40 through 49** (reserved)

Chapters 20 through 39 of the *International Fire Code* (IFC) establish the minimum requirements for special processes and uses. These processes or uses involve a variety of combustible and hazardous materials and are unique to particular applications, such as semiconductor facilities or industrial ovens. For certain activities, such as aviation facilities or motor fuel-dispensing facilities, these processes or uses typically require the storage and handling of thousands of gallons of flammable or combustible liquids. Other chapters address the safe handling and storage of combustible and flammable materials. Chapters 40 through 49 are reserved and serve as placeholders for future code provisions addressing new special occupancies or operations that require regulation by the IFC. Two of those 'future' chapters were added in the 2018 IFC. Chapter 38 for Higher Education Laboratories, and Chapter 39 for Processing and Extraction Facilities. ■

CHAPTER 22

Combustible Dust

2303.2.1

Height of Emergency Disconnect Switch

2306.7.3.1

Protection from Vehicle Impact

2309.6, 2309.6.1

Defueling of Hydrogen Fueled Vehicles

2311.6

Repair of Vehicles Fueled by CNG and LNG

2311.8

Repair of Vehicles Fueled by Lighter-than-air Fuels

2403.2.1.3

Classified Electrical Areas Around Spray Booths

2404.2, 2404.3.1, 914.9

Spray Rooms and Spray Booths

2810

Outdoor Storage of Pallets at Pallet Manufacturing and Recycling Facilities

CHAPTER 31

Umbrella Structures

3103.3.1

Tents and Membrane Structures Used as Special Amusement Buildings

3103.6, 3103.9

Structural Stability of Tents

3104.2

Fabrics for Tents and Membrane Structures

3105, 105.6.47, 105.7.22

Temporary Special Event Structures

3106

Outdoor Assembly Events

3107.13

LP-gas Containers and Tanks Adjacent to Tents and Membrane Structures

CHAPTER 32

High-piled Combustible Storage

3304.5, 3308, 3309.1

Fire Watch During Construction and Demolition

CHAPTER 38

Higher Education Laboratories

CHAPTER 39

Processing and Extraction Facilities

Chapter 22
Combustible Dust

CHANGE TYPE: Modification

CHANGE SUMMARY: Reference to the new NFPA 652, "Standard on the Fundamentals of Combustible Dust," is added to provide guidance and criteria when evaluating combustible dust hazards.

International Code Council®

This sanding operation is creating combustible dust.

2018 CODE:

SECTION 2201 GENERAL

2201.1 Scope. The equipment, processes and operations involving dust explosion hazards shall comply with the provisions of this ~~chapter~~ code and NFPA 652.

SECTION 2202 DEFINITION

2202.1 Definition. The following term is defined in Chapter 2:
COMBUSTIBLE DUST.

SECTION 2203 PRECAUTIONS

2203.1 Owner responsibility. The owner or operator of a facility with operations that manufacture, process, blend, convey, repackage, generate or handle potentially combustible dust or combustible particulate solids shall be responsible for compliance with the provisions of this code and NFPA 652.

2203.2 Dust hazard analysis (DHA). The requirements of NFPA 652 apply to all new and existing facilities and operations with combustible dust hazard. Existing facilities shall have a dust hazard analysis (DHA) completed in accordance with Section 7.1.2 of NFPA 652.

The fire code official shall be authorized to order a dust hazard analysis to occur sooner if a combustible dust hazard has been identified in a facility that has not previously performed an analysis.

~~2203.1~~ 2203.3 Sources of ignition. Smoking, the use of heating or other devices employing an open flame, or the use of spark-producing equipment is prohibited in areas where *combustible dust* is generated, stored, manufactured, processed or handled.

~~2203.2~~ 2203.4 Housekeeping. Accumulation of *combustible dust* shall be kept to a minimum in the interior of buildings. Accumulated *combustible dust* shall be collected by vacuum cleaning or other means that will not place *combustible dust* into suspension in air. Forced air or similar methods shall not be used to remove dust from surfaces.

Chapter 22 continues

Chapter 22 continued

SECTION 2204 ~~EXPLOSION PROTECTION~~ADDITIONAL REQUIREMENTS

2204.1 ~~Standards~~Specific hazards standards. ~~The fire code official is authorized to enforce applicable provisions of the~~ The industry- or commodity-specific codes and standards listed in Table 2204.1 ~~to prevent and control dust explosions~~ shall be complied with based on the identification and evaluation of the specific fire and deflagration hazards that exist at a facility.

TABLE 2204.1 ~~Explosion Protection~~ Specific Hazards Standards

STANDARD	SUBJECT
NFPA 61	Standard for the Prevention of Fires and Dust Explosions in Agricultural and Food Processing Facilities
NFPA 69	Standard on Explosion Prevention Systems
NFPA 70	National Electrical Code
NFPA 85	Boiler and Combustion System Hazards Code
NFPA 120	Standard for Fire Prevention and Control in Coal Mines
NFPA 484	Standard for Combustible Metals
NFPA 654	Standard for Prevention of Fire and Dust Explosions from the Manufacturing, Processing and Handling of Combustible Particulate Solids
NFPA 655	Standard for the Prevention of Sulfur Fires and Explosions
NFPA 664	Standard for the Prevention of Fires and Explosions in Wood Processing and Woodworking Facilities

CHANGE SIGNIFICANCE: IFC Table 5003.1.1(1) and IBC Table 307.1(1) contain Footnote q which requires a technical report and evaluation of the hazard where combustible dusts are generated or used and could create a fire or explosion hazard. The IFC now references NFPA 652, "Standard on the Fundamentals of Combustible Dust," which is a new standard and provides criteria for assessing the hazard of combustible dusts.

NFPA 652 provides a process to conduct a dust hazard analysis and evaluate a facility to quantify the hazard and identify remedial actions and protections levels. If hazards are identified, the standards listed in Table 2204.1 are then complied with for specific protection and mitigation features.

Section 2203.2 states that a dust hazard analysis is required for both new and existing facilities. NFPA 652, Section 7.1.2 is referenced for application at existing facilities. NFPA 652, Section 7.1.2.2 states that for existing facilities a dust hazard analysis must be completed within 3 years, or at the time when modifications occur, whichever occurs first. The IFC also allows the fire code official to require the dust hazard analysis prior to the 3-year time period allowed in NFPA 652 where a specific hazard is identified.

2303.2.1

Height of Emergency Disconnect Switch

CHANGE TYPE: Addition

CHANGE SUMMARY: This new section provides specific height limitations for emergency disconnect switches for fuel dispensing operations.

2018 CODE: **2303.2.1 Height.** The height of the emergency disconnect switch shall be not less than 42 inches (1067 mm) and not more than 48 inches (1372 mm) measured vertically, from the floor level to the activating button.

CHANGE SIGNIFICANCE: This code change specifies a height requirement for emergency disconnect switches at motor vehicle fuel dispensing facilities.

The 2015 IFC requires the emergency disconnect switch to be:

- Clearly identified.
- Readily accessible.
- No closer than 20 feet and no further than 100 feet from outdoor dispensing operations.

This additional criterion specifies the acceptable height of the emergency disconnect switch. The height limitations are identical to the height limitations for a manual fire alarm box in Section 907.4.2.2.

International Code Council®

The emergency disconnect switch must be in a location that is provided with ready access.

2306.7.3.1

Protection from Vehicle Impact

CHANGE TYPE: Addition

CHANGE SUMMARY: The fire code official has the authority to require additional vehicle impact protection at fuel dispensing facilities.

2018 CODE: **2306.7.3.1 Additional impact protection.** The *fire code official* is authorized to require additional impact protection in accordance with Section 312 where dispensing devices are located in areas near parking areas, multiple dispensing devices, highway on- and off-ramps, and other areas where there is a higher potential for vehicle impacts.

CHANGE SIGNIFICANCE: Fuel-dispensing facilities have changed over the past 10-20 years. Many facilities are larger than the previous generations of service stations and there is more on-site traffic because of dispensing operations being combined with mercantile operations. This has led to layouts that do not fit the traditional "wait in line" concept for fueling. The trend to self-service has prompted vehicles navigating through fueling areas, backing up into available fueling positions, and more traffic hazards. In these facilities, the fire code official is now authorized to require additional impact protection.

The IFC provides options on the method of vehicle impact protection. The designer has options of a 6-inch raised curb, guard posts, or other approved equivalent methods. The industry standard is to place impact protection at the end of a row of pumps or provide a 6-inch curb. The 6-inch raised dispenser mounting island provides protection for a parallel vehicle impact and pushes the tire away prior to impact. However, this does not protect against a perpendicular or angled vehicle impact to the island-type dispenser. Impacts from an angular strike are becoming more frequent resulting from jockeying for position.

The new section gives the fire code official the authority to require additional protection, but does not mandate it. There are often designs and layouts where additional protection would not be deemed necessary. The reference to Section 312 provides the flexibility to accept alternative methods of providing vehicle impact protection.

International Code Council®

Fuel dispensing island with additional impact provided adjacent to middle dispensers.

2309.6, 2309.6.1

Defueling of Hydrogen Fueled Vehicles

CHANGE TYPE: Modification

CHANGE SUMMARY: The requirements for repairing vehicles fueled by compressed or liquefied hydrogen gas have been updated to address current technologies and processes.

2018 CODE: **2309.6** ~~**Defueling of hydrogen from fuel storage containers**~~ **Repairs, purging, defueling and discharge.** The repair, purging, defueling or discharge ~~or defueling hydrogen from fuel storage tanks for the purpose of maintenance, cylinder certification, calibration of dispensers or other~~ activities associated with hydrogen motor fuel supply systems and tanks and the installation of the systems shall be in accordance with ~~Sections 2309.6.1 through 2309.6.1.2.4~~ Chapters 53 and 58 and NFPA 2.

> **Exception:** The fuel supply piping from the fuel storage tank to the engine compartment on a motor vehicle or forklift.

~~**2309.6.1 Methods of discharge.** The discharge of hydrogen from fuel storage tanks shall be accomplished through a closed transfer system in accordance with Section 2309.6.1.1 or an approved method of atmospheric venting in accordance with Section 2309.6.1.2.~~

~~**2309.6.1.1**~~ **2309.6.1** ~~**Closed transfer system**~~ **Documented procedure.** A documented procedure that explains the logic sequence for defueling or discharging ~~the storage tank~~ operations shall be maintained on site and shall be provided to the *fire code official* ~~for review and approval~~ upon request. The procedure shall include what actions the operator is required to take in the event of a low-pressure or high-pressure hydrogen release during discharging activity. Schematic design documents shall be ~~provided~~ maintained on site, illustrating the arrangement of piping, regulators and equipment settings. The ~~*construction documents*~~ schematic shall illustrate the piping and regulator arrangement and shall be shown in spatial relation to the location of the vehicle being defueled and, if applicable, to the compressor, storage vessels and emergency shutdown devices.

~~**2309.6.1.2 Atmospheric venting of hydrogen from fuel storage containers.** Where atmospheric venting is used for the discharge of hydrogen from fuel storage tanks, such venting shall be in accordance with Sections 2309.6.1.2.1 through 2309.6.1.2.1.4.~~

~~**2309.6.1.2.1 Defueling equipment.** Equipment used for defueling shall be listed and labeled or *approved* for the intended use.~~

~~**2309.6.1.2.1.1 Manufacturer's equipment required.** Equipment supplied by the manufacturer shall be used to connect the storage tanks to be defueled to the vent pipe system.~~

International Code Council®

This equipment is used to refuel and defuel hydrogen-fueled vehicles.

2309.6, 2309.6.1 continues

2309.6, 2309.6.1 continued

~~**2309.6.1.2.1.2 Vent pipe maximum diameter.** Defueling vent pipes shall have a maximum inside diameter of 1 inch (25 mm).~~

~~**2309.6.1.2.1.3 Maximum flow rate.** The maximum rate of hydrogen flow through the vent pipe system shall not exceed 1,000 cfm at NTP (0.47 m3/s) and shall be controlled by means of the manufacturer's equipment, at low pressure and without adjustment.~~

~~**2309.6.1.2.1.4 Isolated use.** The vent pipe used for defueling shall not be connected to another venting system used for any other purpose.~~

~~**2309.6.1.2.2 Construction documents.** *Construction documents* shall be provided illustrating the defueling system to be utilized. Plan details shall be of sufficient detail and clarity to allow for evaluation of the piping and control systems to be utilized and include the method of support for cylinders, containers or tanks to be used as part of a closed transfer system, the method of grounding and bonding and other requirements specified herein.~~

~~**2309.6.1.2.3 Stability of cylinders, containers and tanks.** A method of rigidly supporting cylinders, containers or tanks used during the closed transfer system discharge or defueling of hydrogen shall be provided. The method shall provide not less than two points of support and shall be designed to resist lateral movement of the receiving cylinder, container or tank. The system shall be designed to resist movement of the receiver based on the highest gas-release velocity through valve orifices at the receiver's rated service pressure and volume. Supporting structures or appurtenances used to support receivers shall be constructed of noncombustible materials in accordance with the *International Building Code*.~~

~~**2309.6.1.2.4 Grounding and bonding.** Cylinders, containers or tanks and piping systems used for defueling shall be bonded and grounded. Structures or appurtenances used for supporting the cylinders, containers or tanks shall be grounded in accordance with NFPA 70. The valve of the vehicle storage tank shall be bonded with the defueling system prior to the commencement of discharge or defueling operations.~~

~~**2309.6.2 Repair of hydrogen piping.** Piping systems containing hydrogen shall not be opened to the atmosphere for repair without first purging the piping with an inert gas to achieve 1-percent hydrogen or less by volume. Defueling operations and exiting purge flow shall be vented in accordance with Section 2309.6.1.2.~~

~~**2309.6.3 Purging.** Each individual manufactured component of a hydrogen generating, compression, storage or dispensing system shall have a label affixed as well as a description in the installation and owner's manuals describing the procedure for purging air from the system during startup, regular maintenance and for purging hydrogen from the system prior to disassembly (to admit air).~~

~~For the interconnecting piping between the individual manufactured components, the pressure rating must be not less than 20 times the absolute pressure present in the piping when any hydrogen meets any air.~~

~~**2309.6.3.1 System purge required.** After installation, repair or maintenance, the hydrogen piping system shall be purged of air in accordance with the manufacturer's procedure for purging air from the system.~~

CHANGE SIGNIFICANCE: The technical language addressing repairs, purging, defueling, and methods of discharge is deleted form the code and a replaced with a reference to NFPA 2, *Hydrogen Technologies Code*.

Section 2309.6.1 specifies that the procedure must be documented. It is anticipated that this procedure would be reviewed at the time of application for a repair garage operational permit as required in Section 105.6.43. The documented procedure detailing how defueling and discharging activities are to be conducted is to be on site and available to the fire code official.

NFPA 2 serves as a comprehensive document addressing construction and operational issues. These defueling and discharging activities and the installation requirements for the necessary equipment are included in NFPA 2. The defueling operation is covered in Section 18.7 of NFPA 2.

2311.6

Repair of Vehicles Fueled by CNG and LNG

CHANGE TYPE: Addition

CHANGE SUMMARY: The requirements for repairing vehicles fueled by compressed or liquefied natural gas have been updated to address current technologies and processes.

2018 CODE: **2311.6 Vehicles powered by liquefied natural gas (LNG) and compressed natural gas (CNG).** LNG vehicles and CNG vehicles shall comply with Sections 2311.6.1 and 2311.6.2, as applicable.

2311.6.1 Liquefied natural gas (LNG). The LNG vehicle fuel system pressure shall be measured and recorded prior to entering the repair facility. The maximum allowable system pressure shall be not more than 170 psig (1172 kPa). Pressure above 170 psig (1172 kPa) shall be reduced by operating the vehicle or limited venting outdoors, as required.

TABLE 2311.6.2 Temperature Compensated Cylinder Pressure Table[a]

GAS TEMPERATURE °F	PRESSURE IN FULL 3,600 PSI CNG CONTAINER, psig
123.6	4,500
120	4,455
110	4,272
100	4,105
90	3,936
80	3,768
70	3,600
60	3,432
50	3,263
40	3,094
30	2,926
20	2,757
10	2,589
0	2,421
-10	2,253
-20	2,086
-30	1,919
-40	1,753

For SI: °C = [(°F) - 32]/1.8, 1 psig = 6.895 kPa.

a. 3,600 psi service pressure calculated from the standard gas composition used to create the gasoline gallon equivalent (GGE).

2311.6.2 Compressed natural gas (CNG). CNG vehicle fuel system pressure and the ambient temperature shall be measured and recorded prior to entering the repair facility. Pressure greater than the indicated maximum pressure in accordance with Table 2311.6.2 shall be reduced by defueling the vehicle.

CHANGE SIGNIFICANCE: Section 2311.6 has been added and includes specific practices to ensure that LNG and CNG fuels are not released during the repair process. LNG warms over time, which increases the fuel system pressure. The LNG storage tank includes a relief valve set at about 220 psig to protect the system. In a typical LNG system, the relief valve resets at a pressure between 205 and 210 psig. Usually the LNG tank pressure increases at about 15 psig per day so reducing the system pressure to 170 psig prior to entering the repair facility allows a two-day period for repairs before any venting takes place.

Based on the SANDIA analysis and report, the credible natural gas releases during LNG and CNG vehicle maintenance operations are limited to about 128,700 BTUs of fuel over a very short time frame. The released natural gas dissipates quickly and is typically ignitable from the point of release to no more than eight feet from that point. These releases do not create an ignitable mixture at the ceiling of the facility because of rapid dispersion of the gas. There has not been any reported release of a full CNG cylinder due to any maintenance operation in the last 35 years of record keeping. In the 1990s there were full cylinder releases due to faulty temperature activated pressure relief devices, but these devices are no longer in use and the releases were not related to maintenance operations. By measuring the system pressure prior to repairing the vehicle, a safe period for repair can be established.

For CNG, the ambient temperature and the cylinder pressure are compared using Table 2311.6.2. If the pressure exceeds the maximum value in the table, then the CNG vehicle should be defueled to the proper pressure prior to entering the facility.

2311.8

Repair of Vehicles Fueled by Lighter-than-air Fuels

CHANGE TYPE: Modification

CHANGE SUMMARY: The requirements for repairing vehicles fueled by compressed or liquefied compressed gas have been updated to address current technologies and processes.

2018 CODE: **2311.8 Repair garages for vehicles fueled by lighter-than-air fuels.** ~~Repair garages~~ The room, motor vehicle repair booth or motor vehicle repair space containing repair garage activities for the conversion ~~and~~ or repair of vehicles that use CNG, LNG, hydrogen or other lighter-than-air motor fuels shall be in accordance with Sections 2311.8 through ~~2311.7.2.3~~ 2311.8.11 in addition to the other requirements of Section 2311. Repair garages for the repair of vehicles that use hydrogen fuel shall be in accordance with this code and NFPA 2.

Exceptions:

1. Repair garages where work is conducted only on vehicles that have been defueled and their systems purged with nitrogen gas, and where standard operating procedures to document and maintain the fueling status throughout repair operations are approved.
2. Repair garages where work is not performed on the fuel system and is limited to exchange of parts and maintenance not requiring open flame or welding on the CNG-, LNG-, hydrogen- or other lighter-than-air-fueled motor vehicle.
3. Repair garages for hydrogen-fueled vehicles where work is not performed on the hydrogen storage tank and is limited to the exchange of parts and maintenance not requiring open flame or welding on the hydrogen-fueled vehicle. During the work, the entire hydrogen fuel system shall contain ~~a quantity that is~~ less than 200 cubic feet (5.6 m^3) of hydrogen.
4. Repair garages for natural-gas-fueled vehicles where work is not being performed on the fuel storage tank, and is limited to

Photo courtesy of Robert Davidson, Davidson Code Concepts, LLC

The louvred vent in the ceiling is activated by the gas detection system in this repair garage for LNG- and CNG-fueled vehicles.

the exchange of parts and maintenance not requiring open flame or welding on the natural-gas-fueled vehicle. During the work, the natural gas in the vehicle fuel tank shall contain a pressure of not more than 250 psi at 70ºF (1724 kPa at 21ºC).

2311.8.1 Preparation of vehicles for repair. For vehicles powered by gaseous fuels, the fuel shutoff valves shall be closed prior to repairing any portion of the vehicle fuel system.

Vehicles powered by gaseous fuels in which the fuel system has been damaged shall be inspected and evaluated for fuel system integrity prior to being brought into the repair garage. The inspection shall include testing of the entire fuel delivery system for leakage.

2311.8.2 Repair garages used for the repair of hydrogen-fueled vehicles. Repair garages used for the repair of hydrogen-fueled vehicles shall be provided with an *approved* exhaust ventilation system in accordance with the *International Mechanical Code* and Chapter 6 of NFPA 2.

2311.8.3 Motor vehicle repair rooms. Motor vehicle repair rooms shall be enclosed with not less than 1-hour fire barriers constructed in accordance with Section 707 of the *International Building Code*, or horizontal assemblies constructed in accordance with Section 711 of the *International Building Code*, or both, with 1-hour rated opening protectives.

2311.8.4 Motor vehicle repair booths. The design and construction of motor vehicle repair booths shall be in accordance with Sections 2311.8.4.1 through 2311.8.4.4.

2311.8.4.1 Construction. Motor vehicle repair booths shall be constructed of approved noncombustible materials. Where walls or ceiling assemblies are constructed of sheet metal, single-skin assemblies shall be not thinner than 0.0478 inch (18 gage) (1.2 mm) and each sheet of double-skin assemblies shall be no thinner than 0.0359 inch (20 gage) (0.9 mm). Structural sections of motor vehicle repair booths shall be sealed in an *approved* manner.

2311.8.4.2 Surfaces. The interior surfaces of motor vehicle repair booths shall be constructed to permit the free passage of exhaust air from all parts of the interior.

2311.8.4.3 Means of egress. *Means of egress* shall be provided in accordance with Chapter 10.

> **Exception:** *Means of egress* doors from premanufactured motor vehicle repair booths shall be not less than 30 inches (762 mm) in width by 80 inches (2032 mm) in height.

2311.8.4.4 Clear space. Motor vehicle repair booths shall be installed so that all parts of the booth are provided with *ready access* for cleaning.

2311.8 continues

2311.8 continued

A clear area of not less than 3 feet (914 mm) wide shall be maintained on all sides of the motor vehicle repair booth. This clear area shall be kept free of any storage or combustible construction.

Exceptions:

1. This requirement shall not prohibit locating a motor vehicle repair booth closer than 3 feet (914 mm) to or directly against an interior partition, wall or floor/ceiling assembly that has a fire-resistance rating of not less than 1 hour, provided the motor vehicle repair booth can be adequately maintained and cleaned.
2. This requirement shall not prohibit locating a motor vehicle repair booth closer than 3 feet (914 mm) to an exterior wall or a roof assembly, provided that the wall or roof is constructed of noncombustible material and the motor vehicle repair booth can be adequately maintained and cleaned.

2311.8.5 Motor vehicle repair spaces. Where such spaces are not separately enclosed, noncombustible spray curtains shall be provided to restrict the spread of flammable gases.

2311.8.6 Fire protection. Motor vehicle repair booths or spaces installed in a room or area protected by an *automatic sprinkler system* shall have the protection extended to include the inside of the motor vehicle repair booth or space.

2311.8.7 Fire extinguishers. Portable fire extinguishers complying with Section 906 shall be provided for motor vehicle repair rooms, motor vehicle repair booths or motor vehicle repair spaces.

~~2311.7.1~~ 2311.8.8 Exhaust ventilation system. Repair garages used for the repair of ~~natural gas- or hydrogen-fueled vehicles~~ CNG, LNG or other lighter-than-air motor fuels other than hydrogen shall be provided with an *approved* mechanical ventilation system. The mechanical ventilation system shall be in accordance with the *International Mechanical Code* and Sections ~~2311.7.1.1~~ 2311.8.8.1 and ~~2311.7.1.2~~ 2311.8.8.2.

Exception: ~~Repair garage with~~ Where *approved* by the *fire code official*, natural ventilation ~~when *approved*~~ shall be permitted in lieu of mechanical exhaust ventilation.

~~2311.7.1.1~~ 2311.8.8.1 Design. ~~Indoor locations shall be ventilated utilizing~~ For indoor locations, air supply inlets and exhaust outlets for mechanical ventilation shall be arranged to provide ~~uniform~~ uniformly distributed air movement ~~to the extent practical. Inlets shall be~~ with inlets uniformly arranged on ~~exterior~~ walls near floor level~~. Outlets shall be located~~ and outlets at the high point of the room in ~~exterior~~ walls or the roof.

~~Ventilation shall be by a continuous mechanical ventilation system or by a mechanical ventilation system activated by a continuously monitoring natural gas detection system or, for hydrogen, a continuously monitoring flammable gas detection system, each activating at a gas concentration~~

~~of not more than 25 percent of the lower flammable limit (LFL). In all cases, the system shall shut down the fueling system in the event of failure of the ventilation system.~~

Failure of the ventilation system shall cause the fueling system to shut down.

The exhaust ventilation rate shall be not less than 1 cubic foot per minute (0.03 m^3/minute) per 12 cubic feet ~~[0.00139 $m^3 \times (s \cdot m^3)$]~~ (34 m^3) of room volume.

~~2311.7.1.2~~ 2311.8.8.2 Operation. The mechanical exhaust ventilation system shall operate continuously.

Exceptions:

1. Mechanical exhaust ventilation systems that are interlocked with a gas detection system designed in accordance with Sections ~~2311.7.2~~ 2311.8.9 through ~~2311.7.2.3~~ 2311.8.9.2.
2. Mechanical exhaust ventilation systems in repair garages that are used only for repair of vehicles fueled by liquid fuels or odorized gases, such as CNG, where the ventilation system is electrically interlocked with the lighting circuit.

~~2311.7.2~~ 2311.8.9 Gas detection system. Repair garages used for repair of vehicles fueled by nonodorized gases including, but not limited to, hydrogen and nonodorized LNG, shall be provided with a ~~flammable~~ gas detection system in accordance with Section 916. The gas detection system shall be designed to detect leakage of nonodorized gaseous fuel. Where lubrication or chassis service pits are provided in garages used for repairing nonodorized LNG-fueled vehicles, gas sensors shall be provided in such pits.

~~**2311.7.2.1 System design.** The flammable gas detection system shall be *listed* or *approved* and shall be calibrated to the types of fuels or gases used by vehicles to be repaired. The gas detection system shall be designed to activate when the level of flammable gas exceeds 25 percent of the lower flammable limit (LFL). Gas detection shall be provided in lubrication or chassis service pits of repair garages used for repairing nonodorized LNG-fueled vehicles.~~

~~**2311.7.2.1.1 Gas detection system components.** Gas detection system control units shall be listed and labeled in accordance with UL 864 or UL 2017. Gas detectors shall be listed and labeled in accordance with UL 2075 for use with the gases and vapors being detected.~~

~~2311.7.2.2~~ 2311.8.9.1 ~~Operation~~ System activation. Activation of the gas detection ~~system~~ alarm shall result in all the following:

1. Initiation of ~~distinct~~ local audible and visual ~~alarm signals~~ alarms in ~~the repair garage~~ approved locations.
2. Deactivation of ~~all~~ heating systems located in the repair garage.
3. Activation of the mechanical exhaust ventilation system, where the ventilation system is interlocked with gas detection.

2311.8 continues

2311.8 continued

~~**2311.7.2.3**~~ **2311.8.9.2 Failure of the gas detection system.** Failure of the gas detection system shall ~~result in the deactivation of~~ automatically deactivate the heating system, ~~activation of~~ activate the mechanical exhaust ventilation system where the system is interlocked with the gas detection system and cause a trouble signal to sound in an *approved* location.

2311.8.10 Classified electrical area. Areas within 18 inches (450 mm) of a ceiling within a motor vehicle repair room or motor vehicle repair booth shall be designed and installed in accordance with the requirements for Class I, Division 2 classified locations, as set forth in NFPA 70.

Exceptions:

1. Rooms with exhaust ventilation of not less than 1 cfm per square foot (0.03 m^3/minute/m^2) of floor area, with suction taken from a point within 18 inches (450 mm) of the highest point in the ceiling in repair garages for vehicles that use CNG, liquefied natural gas (LNG) or other lighter-than-air motor fuels.
2. Rooms used for the repair of hydrogen-fueled vehicles that have an approved exhaust ventilation system in accordance with the *International Mechanical Code* and NFPA 2.

~~**2311.8**~~ **2311.8.11 Defueling equipment required at vehicle maintenance and repair facilities.** Facilities for repairing or replacing hydrogen fuel ~~systems~~ tanks on hydrogen-fueled vehicles shall have equipment to defuel vehicle storage tanks. Where work must be performed on a vehicle's fuel storage tank for the purpose of maintenance, repair or cylinder certification, defueling and purging shall be conducted in accordance with Section 2309.6 and NFPA 2.

CHANGE SIGNIFICANCE: Many revisions have occurred with regard to repair of vehicles fueled by compressed or liquefied gas. Many of these revisions are based on the findings of the SANDIA Report (SAND2014-2342).

Initially, Section 2311.8 has been clarified by adding two exceptions removing certain repair operations from regulation in this section. Exception 1 removes vehicles which have either emptied and purged the fuel tank with nitrogen or the operating procedures monitor the gas pressures on a routine basis. Exception 4 states that if work is not performed on the then fuel tank and pressures are maintained below a maximum of 250 psig, then that repair work is exempt from complying with these requirements.

Section 2311.8.1 has been added and includes specific practices to ensure that LNG and CNG fuels are not released during the repair process. LNG warms over time, which increases the fuel system pressure. The LNG storage tank includes a relief valve set at about 220 psig to protect the system. In a typical LNG system, the relief valve resets at a pressure between 205 and 210 psig. Usually the LNG tank pressure increases at about 15 psig per day; reducing the system pressure to 170 psig prior to entering the repair facility allows a two-day period for repairs before any venting takes place.

Based on the SANDIA analysis and report, the credible natural gas releases during LNG and CNG vehicle maintenance operations are limited to about 128,700 BTUs of fuel over a very short time frame. The released natural gas dissipates quickly and is typically ignitable from the point of release to no more than eight feet from that point. These releases do not create an ignitable mixture at the ceiling of the facility due to rapid dispersion of the gas. There has not been any reported release of a full CNG cylinder due to any maintenance operation in the last 35 years of record keeping. In the 1990s there were full cylinder releases because of faulty temperature activated pressure relief devices, but these devices are no longer in use and the releases were not related to maintenance operations. By measuring the system pressure prior to repairing the vehicle, a safe period for repair can be established. The ambient temperature and the cylinder pressure are compared using Table 2311.6.2. If the pressure exceeds the maximum value in the table, then the CNG vehicle should be defueled to the proper pressure prior to entering the facility. Section 2311.8 Exception 1 allows an approved method of monitoring gas pressures as part of the routine operations in lieu of complying with the mechanical exhaust ventilation requirements.

Where a mechanical exhaust ventilation system is required, the design is based on one of two criteria: 1) if the fuel is hydrogen, the referenced requirements are in NFPA 2 *Hydrogen Technologies Code*; or 2) if the fuel is a lighter-than-air fuel other than hydrogen, the requirements are found in the IFC and IMC. Generally speaking, NFPA 2 requires an exhaust rate of 1 cfm/square foot for hydrogen, and the IFC requires an exhaust rate of 1 cfm for 12 cubic feet of room volume for CNG or LNG.

New Sections 2311.8.3 and 2311.8.4 allow the application of the ventilation and construction requirements to apply to either the room used for repair or the use of a repair booth. The repair booth construction and separation requirements are consistent with the requirement for spray booths. This option allows the owner to design the building to comply with the requirements for lighter-than-air fueled vehicles or simply install a repair booth where the lighter-than-air fueled vehicles are repaired. The repair booth is designed to handle the lighter-than-air fuels.

Section 2311.7.8.1.1 of the 2015 IFC is deleted. This section required that gas detection components be listed for the gas being detected. Currently, there are no hydrogen gas detection systems listed in accordance with this section due to practical difficulties complying with the standards. The requirement for gas detection equipment to be designed for the specific gas being detected is now found in Section 916.3 providing the code official the authority to approve or disapprove detection systems presented for review and approval. For more information on gas detection systems, see Significant Change to Section 916.

2403.2.1.3

Classified Electrical Areas Around Spray Booths

CHANGE TYPE: Modification

CHANGE SUMMARY: The size of the classified area around spray booth openings is reduced to 3 feet.

2018 CODE: **2403.2.1.3 Areas adjacent to spray booths.** Electrical wiring and equipment located outside of, but within ~~5 feet (1524 mm) horizontally and~~ 3 feet (914 mm) ~~vertically~~ of openings in a spray booth or a spray room, shall be *approved* for Class I, Division 2 or Class II, Division 2 hazardous locations, whichever is applicable.

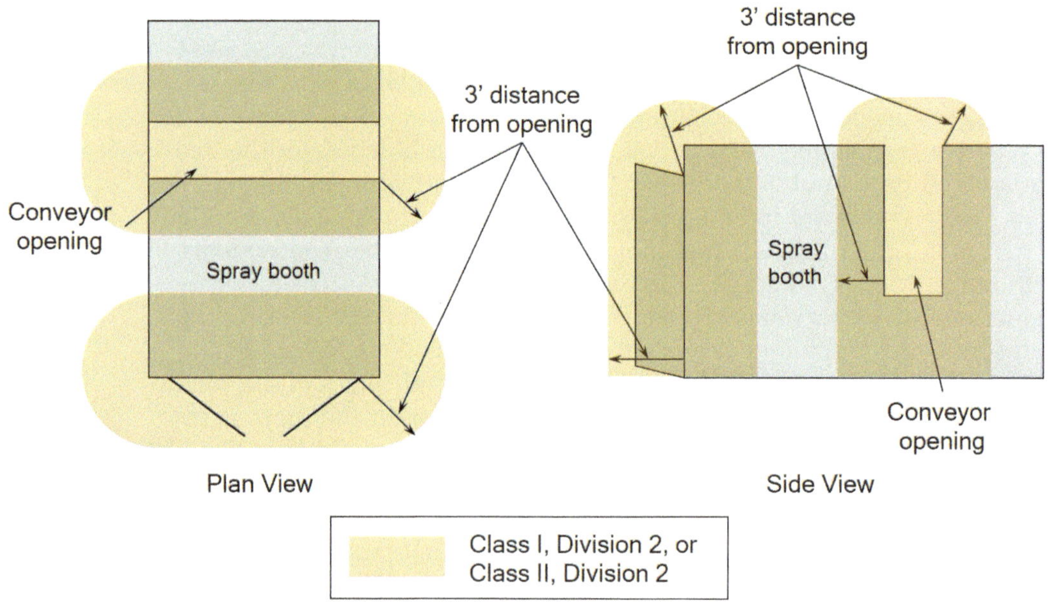

Classified electrical wiring and equipment is required within 3 feet of openings in spray booths.

CHANGE SIGNIFICANCE: The 2015 *International Fire Code* required that any classified (hazardous) location for electrical wiring and equipment was within a distance of 5 feet horizontally from openings in a spray booth. This distance was reduced to 3 feet in both NFPA 70, *National Electrical Code* and NFPA 33, "Standard for Spray Application Using Flammable or Combustible Materials." Therefore, this revision correlates with the referenced standards.

This reduces the size of the area where classified electrical components are required. As a result of the change in the 2015 IFC which requires monitoring of the atmosphere within the spray booth to a maximum of 25 percent of the lower flammable limit, this area should never be at or above 25 percent of the lower flammable limit. Section 2404.7.1 requires the ventilation system to be operating during all spray activities and to continue when flammable vapors are present.

2404.2, 2404.3.1, 914.9

Spray Rooms and Spray Booths

CHANGE TYPE: Modification

CHANGE SUMMARY: Requirements for spray booths and spray operations are correlated between the IFC and the IBC.

2018 IFC: 2404.2 Location of spray-finishing operations. Spray-finishing operations conducted in buildings used for Group A, E, I or R occupancies shall be located in a spray room protected with an *approved automatic sprinkler system* installed in accordance with Section 903.3.1.1 and separated vertically and horizontally from ~~other areas~~ the remainder of the building by *fire barrier* walls and horizontal assemblies with not less than a 1-hour *fire-resistance rating* in accordance with the *International Building Code*. In other occupancies, spray-finishing operations shall be conducted in a spray room, spray booth or limited spraying space *approved* for such use.

Exceptions:

1. Automobile undercoating spray operations and spray-on automotive lining operations conducted in areas with *approved* natural or mechanical ventilation shall be exempt from the provisions of Section 2404 when *approved* and where utilizing Class IIIA or IIIB *combustible* liquids.
2. In buildings other than Group A, E, I or R occupancies, *approved* limited spraying space in accordance with Section 2404.9.
3. Resin application areas used for manufacturing of reinforced plastics complying with Section 2409 shall not be required to be located in a spray room, spray booth or spraying space.

2404.3.1 Spray rooms. Spray rooms shall be constructed and designed in accordance with Section ~~2404.3.1.1 and~~ 416 of the *International Building Code* and Section 2404.3.2 of this code, and shall comply with Sections 2404.4 through 2404.8.2 of this code.

~~2404.3.1.1~~ 2404.3.2 Floor. *No change to text...*

914.9 Application of flammable finishes. An *automatic sprinkler system* or fire-extinguishing system shall be provided in all spray, ~~dip~~ rooms and ~~immersing spaces and storage rooms~~ spray booths, and shall be installed in accordance with Chapter 9.

2404.2, 2404.3.1, 914.9 continues

2404.2, 2404.3.1, 914.9 continued

International Code Council®

Construction of this spray booth is regulated by the *International Fire Code.*

2018 IBC:

SECTION 416 SPRAY APPLICATION OF FLAMMABLE FINISHES

[F] 416.1 General. The provisions of this section shall apply to the construction, installation and use of buildings and structures, or parts thereof, for the spray application of flammable finishes. ~~Such construction~~ Operations and equipment shall comply with the *International Fire Code.*

[F] 416.2.1 Construction. Walls and ceilings of spray rooms shall be constructed of noncombustible materials or the interior surface shall be completely covered with noncombustible materials. Aluminum shall not be used.

[F] ~~416.2.1~~ 416.2.2 Surfaces. The interior surfaces of spray rooms shall be smooth and shall be so constructed to permit the free passage of exhaust air from all parts of the interior and to facilitate washing and cleaning, and shall be so designed to confine residues within the room. ~~Aluminum shall not be used.~~

[F] ~~416.2.2~~ 416.2.3 Ventilation. Mechanical ventilation and interlocks with the spraying operation shall be in accordance with the *International Fire Code* and *International Mechanical Code.*

[F] 416.3.1 Surfaces. The interior surfaces of spraying spaces shall be smooth ~~and continuous without edges~~; shall be so constructed to permit the free passage of exhaust air from all parts of the interior and to facilitate washing and cleaning; and shall be so designed to confine residues within the spraying space. Aluminum shall not be used.

[F] 416.5 Fire protection. An *automatic sprinkler system* or *fire-extinguishing system* shall be provided in all spray~~, dip~~ rooms and ~~immersing spaces and storage rooms~~ spray booths, and shall be installed in accordance with Chapter 9.

SECTION 202 DEFINITIONS

SPRAY ROOM. A room designed to accommodate spraying operations, constructed in accordance with the *International Building Code* ~~and separated from the remainder of the building by a minimum 1-hour fire barrier.~~

CHANGE SIGNIFICANCE: This code change correlates requirements for spray operations in the *International Building Code* and the *International Fire Code*.

The 1-hour separation construction requirement is removed from the definition of spray room and placed into Section 2404.2, which correlates with IBC Section 416.2 which also requires the 1-hour separation.

The term "limited" is added to correlate with the limited spray area requirements in IFC Section 2404.9. A "spraying space" is the result of spraying operation and the boundary of the hazardous vapors; it is not an area to designate as a spraying location. By definition, a "limited spraying space" is an area where a limited amount of spraying can occur outside of a spray room or spray booth.

The term "dip tank" is removed from IBC Section 416.5, so the remaining provisions in IBC Section 416 all apply to spray operations. Therefore, the title of IBC Section 416 is revised to refer to the spray application of flammable finishes. This includes spray rooms, spray booths and spraying spaces.

Dip tanks are regulated as a piece of equipment in the IFC. IFC Section 914.9 and IBC Section 416.5 are revised to be consistent with IFC Section 2405.4. A fire-extinguishing system is not required for all dipping operations. Dip tanks less than 150 gallons can be protected with either a fire-extinguishing system or an automatic closing cover.

Requirements for the construction of spray rooms are also correlated. IFC Section 2404.3.1 is revised to match IBC Section 416.2 for construction of spray rooms. The IFC contains the requirements for the construction of spray booths, since they are treated as an appliance and considered not part of the structure.

2810

Outdoor Storage of Pallets at Pallet Manufacturing and Recycling Facilities

CHANGE TYPE: Addition

CHANGE SUMMARY: This new section adds criteria for outdoor pallet storage at pallet manufacturing facilities and pallet recycling facilities. It provides specific height limits and separation to property lines and buildings, but also allows for the distances to be modified based on providing additional fire protection features.

2018 CODE:

SECTION 2810
OUTDOOR STORAGE OF PALLETS AT PALLET MANUFACTURING AND RECYCLING FACILITIES

2810.1 General. The outside storage of wood pallets and wood composite pallets on the same site as a pallet manufacturing or recycling facility shall comply with Sections 2810.2 through 2810.11.

2810.2 Site plan. Each site shall maintain a current site plan that includes a general description of the property, the boundaries of the lot, the size and location of buildings, and all of the following:

1. Utilities.
2. Type of construction and presence of sprinkler protection for buildings on the site.
3. Water supply sources for fire-fighting purposes.
4. Location of hazardous material storage areas.
5. Location of pallet storage.
6. Equipment protected with a dust collection system.
7. *Fire apparatus access roads.*
8. Designated smoking areas.
9. Location of fire alarm control panels.

2810.3 Fire prevention plan. The *owner* or *owner's* authorized representative shall prepare an *approved* fire prevention plan that includes all of the following:

1. Frequency of walk-through inspections to verify compliance with the plan.
2. Hot work permit program in accordance with Chapter 35.
3. Preventive maintenance program for equipment associated with pallet activities.
4. Inspection, testing and maintenance of *fire protection systems* in accordance with Chapter 9.

2810.4 Fire safety and evacuation plan. The *owner* or *owner's* authorized representative shall prepare and train employees in an *approved* fire safety and emergency evacuation plan in accordance with Chapter 4.

2810.5 Security management plan. The *owner* or *owner's* authorized representative shall prepare a security management plan based on a security risk assessment and shall make the plan and assessment available to the *fire code official* upon request.

2810.6 Clearance to property line. Stacks of pallets shall not be stored within 0.75 times the stack height or 8 feet (2438 mm) of the property line, whichever is greater, or shall comply with Section 2810.11.

2810.7 Clearance to important buildings. Stacks of pallets shall not be stored within 0.75 times the stack height of any important building on site, or shall comply with Section 2810.11.

2810.8 Height. Pallet stacks shall not exceed 20 feet (6096 mm) in height.

2810.9 Fire flow. Fire flow requirements for the site shall be determined by the *fire code official.*

2810.10 Portable fire extinguishers. Portable fire extinguishers shall be provided with 75 feet (22 860 mm) of any pallet stack.

International Code Council®

Pallet storage at pallet recycling facilities is limited to 20 feet in height. Separation to property line is 75 percent of the stack height but can be reduced based on additional fire protection features.

2810.11 Alternative approach. Where *approved* by the *fire code official*, pallet stacks located closer to a property line or structure than as required by Sections 2810.6 and 2810.7 shall be provided with additional fire protection including, but not limited to, the following:

1. The storage yard areas and materials-handling equipment selection, design and arrangement are based on an *approved* risk assessment.

2810 continues

2810 continued

2. Automatic fire detection which transmits an alarm to a supervising station in accordance with NFPA 72.

3. *Fire apparatus access roads* around all storage areas.

CHANGE SIGNIFICANCE: Outdoor pallet storage areas at pallet manufacturing and pallet recycling operations are constantly in a state of flux. Pallets are being moved and replaced daily. As a result, the outdoor pallet storage area is an active management environment. Pallet manufacturers and recyclers have intimate knowledge of their pallet inventory, as it is considered an asset. Personnel are a constant presence within the storage area so that fire hazards can be identified and reported to take immediate corrective action.

This new section applies specifically to pallet manufacturing and pallet recycling facilities. The intent of the new section is to reduce the likelihood of fire at pallet manufacturing and recycling facilities through best practices.

Pallet storage is limited to 20 feet in height, but there is no limit to the footprint of the pile. Separations to property lines are specified at 75 percent of the pile height, or a minimum of 8 feet. Separation to important buildings is specified at 75 percent of the pile height. Note that the separation distance to buildings is different for outdoor pallet storage than the distance found in the IFC Section 315.7.6. (See Significant Change to Section 315.7.) This section is specific for pallet manufacturing and pallet recycling, whereas Section 315.7 would apply to any industrial or commercial facility with pallet storage.

It should be noted that the separation to buildings is applicable to "important buildings." Oftentimes, small tool shacks or work areas are constructed within the pallet storage area. These typically small structures provide shade and protection from the weather for workers out in the pallet area. These structures would not be considered an important building. The term "important building" is used in several places in the code and it refers to buildings which are critical to the continued operation or of significant importance to the business.

Section 2810.11 allows the separation distances to be reduced where additional fire safety features are provided. The way that this section is worded, it requires Items 1, 2 and 3, but does not restrict the fire code official from requiring additional protection measures. The required items are 1) risk assessment, 2) fire detection system, and 3) fire apparatus access roads around the perimeter of the storage areas. The risk assessment could address issues such as storage height or separation of piles as part of the fire protection measures.

Section 2810.4 requires a fire safety and evacuation plan. This plan will contain information on pallet storage arrangements, required separations to buildings and property lines, fire apparatus access roads in addition to reporting fire, evacuation and fighting fire. A companion change occurred in Chapter 4 to correlate the requirement for a fire safety and evacuation plan at pallet manufacturing and recycling facilities.

See also Significant Change to Section 315.7 for general outdoor pallet storage.

Chapter 31

Umbrella Structures

CHANGE TYPE: Clarification

CHANGE SUMMARY: A new definition is added for umbrella structures which results in regulation of umbrella structures when they exceed 400 square feet.

2018 CODE:

SECTION 202 DEFINITIONS

TENT. A structure, enclosure, umbrella structure or shelter, with or without sidewalls or drops, constructed of fabric or pliable material supported by any manner except by air or the contents it protects (see "Umbrella structure").

UMBRELLA STRUCTURE. A structure, enclosure or shelter with or without sidewalls or drops, constructed of fabric or pliable material supported by a central pole or poles (see "Tent").

International Code Council®

This umbrella structure is regulated as a tent.

CHANGE SIGNIFICANCE: Umbrella structures are available that meet or exceed an individual size of 400 square feet. This new definition will include umbrella structures under the definition of "Tent." Therefore, wherever the term tent is used, it now specifically includes umbrella structures. Essentially, an umbrella structure is a tent with one center pole, or multiple center poles, rather than perimeter support poles.

Umbrella structures are available with optional accessories that allow multiple umbrellas to be connected to each other for additional size and stability which can offer extended weather protection. These accessories include rain gutter systems, side panels (with or without support poles), and additional electrical wiring for lights and heater attachments. These accessories allow the umbrella structures to mimic a tent without being considered a "tent" under the conventional use of the term "Tent."

Umbrellas are typically not regulated because of their singular use and limited size. But, when connected or attached, they represent the same hazards as a tent or membrane structure.

The result of these changes is:

- Fabric must meet the flame-spread characteristics required for tents.
- A permit is required for single umbrella structures over 400 square feet, or when the aggregate of multiple umbrella structures results in an area over 700 square feet.
- Umbrella structures must be separated from buildings and property lines.
- Umbrella structures must be structurally sound.

3103.3.1

Tents and Membrane Structures Used as Special Amusement Buildings

CHANGE TYPE: Addition

CHANGE SUMMARY: Special amusement buildings located in temporary tents are required to be equipped with an automatic sprinkler system.

2018 CODE: **3103.3.1 Special amusement building.** Tents and other membrane structures erected as a special amusement building shall be equipped with an *automatic sprinkler system* in accordance with Section 411.3 of the *International Building Code.*

Photo courtesy of Ed Kaminski, Clark County Building and Fire Prevention Services

This special amusement building is located in a temporary membrane structure.

CHANGE SIGNIFICANCE: Section 411 of the IBC contains requirements for Special Amusement Buildings since they present unique egress challenges. Section 411.3 requires that an automatic sprinkler system be installed in all special amusement buildings when they are 1,000 square feet (93 m^2) or greater in area, or have an exit access travel distance of 50 feet (15 240 mm) or more.

A special amusement building regulated in the IFC is a building that is:

- Temporary, permanent or mobile
- Contains a device or system that conveys passengers or provides a walkway along, around or over a course in any direction as a form of amusement
- Arranged so that the egress path is not readily apparent due to visual or audio distractions or an intentionally confounded egress path, or is not readily available because of the mode of conveyance through the building or structure.

Tents and membrane structures have become a popular option for temporary special amusement buildings. Temporary membrane structures require an operational permit in IFC Section 105.6.47. The code anticipates temporary special amusement buildings because it references a temporary water supply with regard to these temporary uses.

The fact that the special amusement building is temporary does not eliminate or reduce the inherent hazards in these facilities. This section increases occupant safety by requiring a fire sprinkler system.

Note that IBC Section 411.3 can also be found in the IFC in Section 914.7.1. The requirements are the same, and both the IBC and IFC Section 914.7.1 apply to both temporary and permanent construction. The new Section 3103.3.1 specifically requires a fire sprinkler system even though the special amusement buildings is temporary. IBC Section 411.3 and IFC Section 914.7.1 allow a temporary water supply when the special amusement building is temporary.

3103.6, 3103.9

Structural Stability of Tents

CHANGE TYPE: Modification

CHANGE SUMMARY: Temporary tents and membrane structures are required to provide construction documents which address their structural stability and load carrying capacity. Larger tents and membrane structures have been added to the list of temporary facilities which must comply.

2018 CODE: **3103.6 Construction documents.** A detailed site and floor plan for tents or membrane structures with an *occupant load* of 50 or more shall be provided with each application for approval. The tent or membrane structure floor plan shall indicate details of the *means of egress* facilities, seating capacity, arrangement of the seating and location and type of heating and electrical equipment. The *construction documents* shall include an analysis of structural stability.

3103.9 ~~Anchorage~~ Structural stability and anchorage required. Tents or membrane structures and their appurtenances shall be ~~adequately roped, braced and anchored~~ designed and installed to withstand the elements of weather and prevent collapsing. Documentation of structural stability shall be furnished to the *fire code official.* ~~on request.~~

3103.9.1 Tents and membrane structures ~~exceeding~~ greater than one story. Tents and membrane structures exceeding one story shall be designed and constructed to comply with ~~Chapter 16~~ Sections 1606 through 1609 of the *International Building Code.*

3103.9.2 Tents and membrane structures greater than 7,500 square feet. *Tents* and *membrane structures* greater than 7,500 square feet (697 m^2) shall be designed and constructed to comply with Sections 1606 through 1609 of the *International Building Code.*

3103.9.3 Tents and membrane structures with an occupant load greater than 1,000. *Tents* and *membrane structures* with an occupant capacity greater than 1,000 persons shall be designed and constructed to comply with Sections 1606 through 1609 of the *International Building Code.*

CHANGE SIGNIFICANCE: This code change addresses the structural stability of temporary tents and membrane structures. This code change was developed by a Fire Code Action Committee Work Group comprised of code officials, industry representatives and other stakeholders.

Initially, Section 3103.6 requires that the construction documents must include an analysis of the structural stability. In correlation with this requirement, Section 3103.9 states that the documentation be provided to the fire code official.

The two new sections added to the IFC require evaluation under specific sections in the IBC. The 2015 IFC required compliance with the structural criteria when the temporary tent or membrane structure has more than one story. In addition to tents and membrane structures greater than one story, temporary tents and membrane structures greater than 7,500 square feet or with an occupant load greater than 1,000 are now required to provide the same structural analysis, even though they may be only one story.

Photo courtesy of Assistant Fire Marshal Fulton Cochran, Clark County Department of Building and Fire Prevention

This large tent, with a floor area over 7,500 square feet, must provide an analysis of its structural integrity.

The 2015 IFC contained a reference in Section 3103.9.1 stating that tents and membrane structures exceeding one story must comply with IBC Chapter 16. This general reference to Chapter 16 has been replaced with references to specific sections. The temporary tents and membrane structures regulated under the IFC do not need to comply with all of Chapter 16, rather only certain sections. Those sections are:

1. IBC Section 1606—Dead Loads
2. IBC Section 1607—Live Loads
3. IBC Section 1608—Snow Loads
4. IBC Section 1609—Wind Loads

There are other sections in IBC Chapter 16 which are applicable to permanent tents and membrane structures, but the temporary use of these of facilities only need comply with these four sections. These four sections deal with the hazards that have repeatedly caused collapse or problems of these temporary structures.

3104.2

Fabrics for Tents and Membrane Structures

CHANGE TYPE: Addition

CHANGE SUMMARY: The application of testing criteria for flame spread of tent and membrane structures has been clarified.

2018 CODE: **3104.2 Flame propagation performance treatment.** Before a permit is granted, the *owner* or agent shall file with the *fire code official* a certificate executed by an *approved* testing laboratory. ~~certifying that the tents and~~ The certificate shall indicate that the floor coverings, tents, membrane structures and their appurtenances, which include side-walls, drops and tarpaulins, ~~floor coverings,~~ are composed of materials meeting the flame propagation performance of Test Method 2 of NFPA 701 ~~and~~. Additionally, it shall indicate that the bunting and combustible decorative materials and effects~~, including sawdust where used on floors or passageways,~~ are composed of material meeting the flame propagation performance criteria of Test Method 1 or Test Method 2 of NFPA 701 ~~or~~, as applicable. Alternatively, the materials shall be treated with a flame retardant in an *approved* manner and meet the flame propagation performance criteria of the applicable test method ~~1 or Test Method 2, as appropriate,~~ of NFPA 701~~, and that such~~. The flame propagation performance criteria ~~are~~ shall be effective for the period specified by the permit.

CHANGE SIGNIFICANCE: The scope and applicability of NFPA 701, "Standard Methods of Fire Tests for Flame Propagation of Textiles and Films" explains when Test Method 1 is applicable and when Test Method 2 should be used. Test Method 1 applies to materials with densities of 21 ~~oz/yd^2~~ ounces per square yard or less. This includes items such as:

- Fabrics and materials used in curtains, draperies or window treatments
- Single-layer fabrics

Photo courtesy of Assistant Fire Marshal Fulton Cochran, Clark County Department of Building and Fire Prevention

The tent fabric must be tested in accordance with Test Method 2 of NFPA 701.

- ~~Multi-layer~~ Multilayer curtain and drapery assemblies which are fastened together
- Fabrics used in construction applications

Test Method 2 applies to the following:

- Decorative materials other than fabrics specified for Test Method 1
- Fabrics, including multilayered fabrics, films and plastic blinds, with or without reinforcement or backing, with material densities greater than ~~21 oz/yd²~~ 21 ounces per square yard (700 g/m^2)
- Vinyl-coated fabric blackout linings and lined draperies using a vinyl-coated fabric blackout lining
- Plastic films
- Fabrics, with or without reinforcement or backing, used for decorative or other purposes inside a building or as temporary or permanent enclosures for buildings under construction
- Fabrics used in the assembly of awnings, tents, tarps, membrane structures or banners
- Fabrics used in the assembly of awnings, tents, tarps, and similar architectural fabric structures and banners

These revisions apply the appropriate test methodology to the material.

Additionally, the reference to testing sawdust used on the floor is removed from this section. Sawdust cannot be tested in accordance with NFPA 701.

3105, 105.6.47, 105.7.22

Temporary Special Event Structures

International Code Council®

This special event is conducted under a temporary special event structure.

CHANGE TYPE: Addition

CHANGE SUMMARY: The requirements for temporary stage structures are expanded to include all temporary structures greater than 400 square feet when used at special events.

2018 CODE: The term "Temporary stage canopy" has been replaced throughout the code with the term "Temporary special event structure." The revised sections with only a change in terminology are not shown here. The sections which present a change in the application of the code are included below. For the remaining text, see the 2018 *International Fire Code.*

SECTION 3105
TEMPORARY ~~STAGE CANOPIES~~ SPECIAL EVENT STRUCTURES

3105.4 Use period. *Temporary* ~~*stage canopies*~~ *special event structures* erected in accordance with ANSI E1.21 shall not be erected for a period of more than ~~45 days~~ six consecutive weeks.

3105.5 Required documents. The following documents shall be submitted to the *fire code official* and the building official for review before a permit is *approved*:

1. Construction documents: *Construction documents* shall be prepared by a *registered design professional* in accordance with the *International Building Code* ~~by a registered design professional~~ and ANSI E1.21 where applicable. *Construction documents* shall include:
   - 1.1. A summary sheet showing the building code used, design criteria, loads and support reactions.
   - 1.2. Detailed construction and installation drawings.
   - 1.3. Design calculations.
   - 1.4. Operating limits of the structure explicitly outlined by the *registered design professional* including environmental conditions and physical forces.
   - 1.5. Effects of additive elements such as video walls, supported scenery, audio equipment, vertical and horizontal coverings.
   - 1.6. Means for adequate stability including specific requirements for guying and cross-bracing, ground anchors or ballast for different ground conditions.
2. Designation of responsible party: The *owner* of the temporary ~~stage canopy~~ special event structure shall designate in writing a person to have responsibility for the temporary ~~stage canopy~~ special event structure on the site. The designated person shall have sufficient

knowledge of the construction documents, manufacturer's recommendations and operations plan to make judgments regarding the structure's safety and to coordinate with the *fire code official.*

3. Operations plan: The operations plan shall reflect manufacturer's operational guidelines, procedures for environmental monitoring and actions to be taken under specified conditions consistent with the *construction documents.*

3105.9 Portable fire extinguishers. Portable fire extinguishers shall be provided as required by Section 906.

SECTION 105 PERMITS

105.6.47 Temporary membrane structures and tents. An operational permit is required to operate an air-supported temporary membrane structure, a temporary ~~stage canopy~~ *special event structure* or a tent having an area in excess of 400 square feet (37 m^2).

Exceptions:

1. Tents used exclusively for recreational camping purposes.
2. Tents open on all sides, which comply with all of the following:
   2.1. Individual tents having a maximum size of 700 square feet (65 m^2).
   2.2. The aggregate area of multiple tents placed side by side without a fire break clearance of not less than 12 feet (3658 mm) shall not exceed 700 square feet (65 m^2) total.
   2.3. A minimum clearance of 12 feet (3658 mm) to structures and other tents shall be provided.

105.7.22 Special event structure. A single construction permit is required to erect and take down a *temporary special event structure.*

SECTION 202 DEFINITIONS

~~**TEMPORARY STAGE CANOPY.** A temporary ground-supported membrane-covered frame structure used to cover stage areas and support equipment in the production of outdoor entertainment events.~~

TEMPORARY SPECIAL EVENT STRUCTURE. Any temporary ground-supported structure, platform, stage, stage scaffolding or rigging, canopy, tower supporting audio or visual effects equipment or similar structures not regulated within the scope of the *International Building Code.*

3105, 105.6.47, 105.7.22 continues

3105, 105.6.47, 105.7.22
continued

CHANGE SIGNIFICANCE: This code change replaces the term "Temporary stage canopy" with "Temporary special event structure" throughout Chapter 31 and also in Section 105.6.47 for the operational permit. Additionally, a new construction permit is added for assembly or dismantling of a temporary special event structure. To make this transition complete, the definition of "temporary stage canopy" is replaced with a new definition of "temporary special event structures."

The definition for temporary special event structure includes what the 2015 IFC defined as a temporary stage canopy. So, the requirements for temporary stage canopies still exist in the code and now are applicable to other temporary special event structures, which includes all structures related to temporary events used in outdoor situations with an area greater than 400 square feet.

The change will ensure that structures covered by ANSI E1.21, "Entertainment Technology: Temporary Ground Supported Overhead Structures Used to Cover the Stage Areas and Support Equipment in the Production of Outdoor Entertainment Events" will be structurally engineered. This change also ensures that all temporary structures not regulated by the IBC will be regulated by this chapter of the IFC. The addition of the construction permit requirement will ensure that all temporary special event structures will be subject to a plan review and site inspection to ensure structural stability and reliability for public safety.

Section 3105.9 is added to require a fire extinguisher at or near all special event structures. This is consistent with the existing requirement in this chapter for providing fire extinguishers in tents and membrane structures.

3106

Outdoor Assembly Events

CHANGE TYPE: Addition

CHANGE SUMMARY: This section adds requirements specific to outdoor public gatherings and improves the correlation of requirements in the IBC and IFC.

2018 CODE:

SECTION 3106
OUTDOOR ASSEMBLY EVENTS

3106.1 Scope. All outdoor assembly events shall comply with this section.

3106.2 General. Outdoor assembly events shall be in accordance with this section and Section 403.12. Temporary structures erected for outdoor assembly events shall comply with this chapter.

3106.2.1 Approval required. Outdoor assembly events shall be *approved* by the *fire code official*.

3106.2.2 Permits. An operational permit shall be required as set forth in Section 105.6.

3106.2.3 Access. An *approved* means of fire apparatus access shall be provided.

3106.2.3.1 Fire service features. Unobstructed access to fire hydrants, drafting sources and other fire protection features shall be maintained at all times.

3106 continues

International Code Council®

An outdoor event where vehicular access and access to fire protection equipment is impeded.

3106 continued

3106.3 Occupancy and means of egress. The number and location of emergency egress and escape routes shall be *approved* by the *fire code official.*

3106.3.1 Occupant load. The *fire code official* shall establish an *occupant load* for the event site.

3106.3.2 Maintenance of emergency egress and escape routes. Emergency egress and escape routes shall be maintained at all times.

3106.4 Public safety for events. Outdoor assembly events shall comply with Sections 3106.4.1 through 3106.4.7.

3106.4.1 Public safety plan for gatherings. A public safety plan shall be prepared where required by Section 403.12.2. The public safety plan shall be submitted to the *fire code official* with the application for an operational permit as required by Section 3106.2.2.

3106.4.2 Weather monitoring person. Where required by the *fire code official*, the event operator or agent shall designate one qualified individual to continuously monitor local weather reports, forecasts and conditions. Said person shall be responsible for initiating weather-related event mitigation activities, ordering the suspension or cancellation of the outdoor assembly event and issuing the evacuation signal in accordance with the *approved* public safety plan.

3106.4.3 Crowd managers. Where events involve a gathering of more than 1,000 people, trained crowd managers shall be provided in accordance with Section 403.12.3.

3106.4.4 Portable fire extinguishers. *Approved* portable fire extinguishers complying with Section 906 shall be provided and placed in locations *approved* by the *fire code official.*

3106.4.5 Smoking. Smoking shall be permitted only in designated areas. Other areas shall have *approved* "No Smoking" signs conspicuously posted and maintained in accordance with Section 310.

3106.4.6 Combustible vegetation. Combustible vegetation that could create a fire hazard shall be removed from the outdoor assembly event area.

3106.4.7 Combustible refuse. Combustible refuse shall be kept in noncombustible containers with tight-fitting or self-closing lids. Combustible refuse shall be removed from the event site at regular intervals to prevent an unsafe accumulation within the event site.

3106.5 Cooking appliances or devices. Outdoor assembly events with concession stands or booths using cooking appliances or devices shall comply with Sections 3106.5.1 through 3106.5.3.

3106.5.1 Separation from tents or structures. Cooking appliances or devices that produce sparks or grease-laden vapors or flying embers (firebrands) shall not be performed within 20 feet (6096 mm) of a tent or temporary structure.

Exceptions:

1. Designated cooking tents not occupied by the public when *approved* by the *fire code official.*
2. *Tents* or structures where cooking appliances are protected with an automatic fire-extinguishing system in accordance with Section 904.12.

3106.5.2 Protection. Cooking equipment using combustible oils or solids shall meet the following:

1. A noncombustible lid shall be immediately available. The lid shall be of sufficient size to cover the cooking well completely.
2. The equipment shall be placed on a noncombustible surface.
3. An *approved* portable fire extinguisher for the protection of cooking grease fires shall be provided at a location *approved* by the *fire code official.*

3106.5.3 Liquefied petroleum gas (LP-gas). The use of liquefied petroleum gas (LP-gas) shall be in accordance with Chapter 61.

3106.6 Electrical equipment and wiring. Outdoor assembly events with concession stands or booths using electrical equipment and temporary wiring for electrical power or lighting shall comply with the applicable provisions of NFPA 70 and Sections 3106.6.1 through 3106.6.3.

3106.6.1 Outdoor use. Electrical equipment and wiring shall be *listed* and *labeled* for outdoor use.

3106.6.2 Generators. Generators shall be installed not less than 10 feet (3048 mm) from combustible materials, and shall be isolated from the public by physical guard, fence or enclosure installed not less than 3 feet (914 mm) away from the internal combustion power source.

3106.6.3 Portable fire extinguishers. Each generator shall be provided with an *approved* portable fire extinguisher complying with Section 906.

CHANGE SIGNIFICANCE: This section deals specifically with outdoor assemblies and was prompted by several fatal events related to outdoor public gatherings where structural failures have occurred. Events include:

1. The Indiana State Fair stage collapse – August 13, 2011
2. Wood Dale Prairie Fest (Chicago area) – August 2, 2015
3. Lancaster Fairgrounds circus tent collapse (New Hampshire) – August 3, 2015

3106 continues

3106 continued

This proposal adds a new definition for "outdoor assembly event" and it requires an operational permit when the occupant load exceeds 1,000 persons. These new requirements include providing for fire apparatus access, access to and availability of fire protection equipment, maintenance of emergency egress routes and maximum occupant load.

These new requirements also correlate the existing provisions in the 2015 IFC Section 403.12 which addresses public safety at gatherings. Section 403.12 allows the fire code official to require a public safety plan, fire watch and crowd managers for these outdoor events. An additional requirement is added for a weather monitoring person. The fire code official can require a weather monitoring person where deemed necessary. This person shall constantly monitor the weather to determine if conditions are changing which would necessitate the occupants to take cover, evacuate a temporary tent or seek refuge out of harm's way.

Requirements for concession stands and food booths are added to ensure a minimum level of safety requirements for fire and public safety of all types and sizes of concession stands. Cooking appliances generating grease-laden vapors or flying embers must be located outside and at least 20 feet from a tent or temporary structure. However, the distance can be reduced, or the cooking can occur inside a tent if the tent is not occupied by the public or where the tent is protected with an automatic sprinkler system.

Cooking appliances using combustible oils or solid fuel have additional requirements. These cooking appliances must have a noncombustible and tight-fitting lid, must be placed on a noncombustible surface, and must have a properly rated portable fire extinguisher at an approved location.

3107.13

LP-gas Containers and Tanks Adjacent to Tents and Membrane Structures

CHANGE TYPE: Modification

CHANGE SUMMARY: Requirements for the use and separation of LP-gas containers in and around tents and membrane structures have been revised.

2018 CODE: ~~**3104.16**~~ **3107.13 LP-gas.** The storage, handling and use of LP-gas and LP-gas equipment shall be in accordance with Sections ~~3104.16.1~~ 3107.13.1 through ~~3104.16.3~~ 3107.13.3.

~~**3104.16.1**~~ **3107.13.1 General.** LP-gas equipment such as containers, tanks, piping, hoses, fittings, valves, tubing and other related components shall be *approved* and in accordance with Chapter 61 and with the *International Fuel Gas Code.*

~~**3104.16.2**~~ **3107.13.2 Location of containers.** LP-gas containers and tanks shall be located outside in accordance with Table 6104.3. ~~Safety release valves~~ Pressure relief devices shall be pointed away from the *tent* or *membrane structure.*

~~**3104.16.2.1 Containers 500 gallons or less.** Portable LP-gas containers with a capacity of 500 gallons (1893 L) or less shall have a minimum separation between the container and structure not less than 10 feet (3048 mm).~~

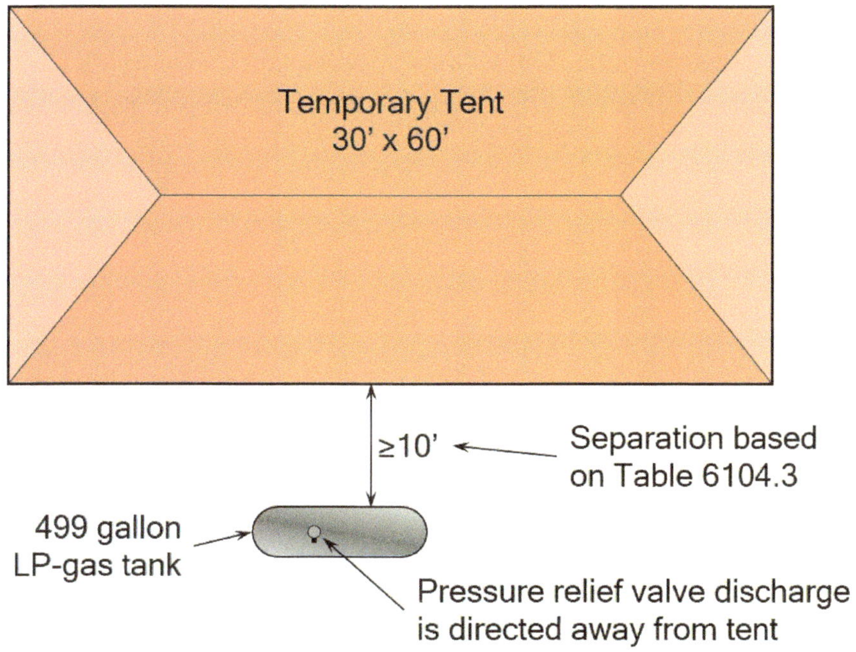

LP-gas storage tank separated from a tent or membrane structure.

3107.13 continues

3107.13 continued

~~**3104.16.2.2 Containers more than 500 gallons.** Portable LP-gas containers with a capacity of more than 500 gallons (1893 L) shall have a minimum separation between the container and structures not less than 25 feet (7620 mm).~~

~~**3104.16.3**~~ **3107.13.3 Protection and security.** Portable LP-gas containers, tanks, piping, valves and fittings that are located outside and are being used to fuel equipment inside a tent or membrane structure shall be adequately protected to prevent tampering, damage by vehicles or other hazards and shall be located in an *approved* location. Portable LP-gas containers shall be ~~securely fastened in place~~ secured to prevent unauthorized movement.

CHANGE SIGNIFICANCE: This revision starts out by relocating the LP-gas requirements from Section 3104 addressing temporary tents and membrane structures, to Section 3107 which covers operational requirements in all situations governed by Chapter 31. In several sections, the term LP-gas "tanks" is incorporated in addition to containers. Based on the definitions in Chapter 2, the term "containers" is only valid up to 60 gallons, and LP-gas containers, including cylinders, can easily exceed that volume.

Section 3107.13.2 is revised by including a reference to Table 6104.3 which addresses LP-gas containers outside of buildings. Since Table 6104.3 is referenced, Sections 3104.16.2.1 and 3104.16.2.2 are deleted as they provided the same information as in the table. There is no technical reason to treat a fabric tent any different than a conventional structure from the standpoint of fire safety.

Another revision occurs in Section 3107.13.3. The last sentence is revised to require that containers and tanks are secured to prevent unauthorized movement. Solutions could include chain and lock to a fixed object, or located within a fenced area. Large containers are typically not a concern because of their own weight.

Chapter 32

High-piled Combustible Storage

CHANGE TYPE: Modification

CHANGE SUMMARY: The requirements in the chapter have been updated to correlate with current NFPA 13 requirements and recent FM Global fire tests.

2018 CODE: Due to the extent of the code text, an outline of the revised sections is shown below. For the complete text, see the 2018 *International Fire Code.*

3201 GENERAL

3201.1 Scope.

3201.2 Permits.

3201.3 Construction documents.

3201.3.1 Approved construction documents.

3201.3.2 Approved storage layout.

3201.4 ~~Evacuation~~ Fire safety and evacuation plan.

3202 DEFINITIONS *....no change*

3203 COMMODITY CLASSIFICATION

3203.1 Classification of commodities.

3203.2 Class I commodities.

3203.3 Class II commodities.

3203.4 Class III commodities.

3203.5 Class IV commodities.

3203.6 High-hazard commodities.

3203.7 Classification of plastics.

3203.7.1 Group A plastics.

3203.7.2 Group B plastics.

3203.7.3 Group C plastics.

3203.8 Examples of commodity classification.

~~3203.7.4~~ 3203.9 Limited quantities of Group A plastics in mixed commodities.

3203.9.1 Classifying mixed commodities with limited Group A plastics.

3203.9.2 Percentage of Group A plastics.

3203.10 Plastic pallets.

3203.10.1 Unreinforced plastic pallets.

3203.10.2 Reinforced plastic pallets.

3203.10.3 Other pallets.

Chapter 32 continues

Chapter 32 continued

3204 DESIGNATION OF HIGH-PILED STORAGE AREAS

3204.1 General.

3204.2 Designation based on engineering analysis.

3205 HOUSEKEEPING AND MAINTENANCE*....no change*

3206 GENERAL FIRE PROTECTION AND LIFE SAFETY FEATURES

3206.1 General.

3206.2 ~~Extent and type~~ Type of protection.

3206.2.1 Extent of protection.

Table 3206.2 General Fire Protection and Life Safety Requirements

3206.3 ~~Separation of high-piled~~ High-piled storage areas.

3206.3.1 Size of high-piled storage area.

3206.3.2 Multiple high-piled storage areas.

~~3206.3.2.2~~ 3206.3.2.1 ~~Multiclass~~ Multiple class high-piled storage areas.

3206.4 Automatic sprinklers.

3206.4.1 Pallets.

3206.4.1.1 Plastic pallets.

3206.5 Fire detection.

3206.6 Building access.

~~3206.6.1~~ 3206.7 ~~Access~~ Fire department access doors.

3206.7.1 Exterior walls without fire department access doors.

3206.7.2 Where located.

3206.7.3 Access to doors.

3206.7.4 Marking on fire department access doors.

~~3206.6.1.1~~ 3206.7.5 Number of doors required.

~~3206.6.1.2~~ 3206.7.6 Door size and type.

~~3206.6.1.3~~ 3206.7.7 Locking devices.

3206.7.8 Key box.

~~3206.7~~ 3206.8 Smoke and heat removal.

~~3206.8~~ 3206.9 Fire department hose connections.

~~3206.9~~ 3206.10 Aisles.

~~3206.9.1~~ 3206.10.1 Width.

~~3206.9.1.1~~ 3206.10.1.1 Sprinklered buildings.

~~3206.9.1.2~~ 3206.10.1.2 Nonsprinklered buildings.

~~3206.9.2~~ 3206.10.2 Clear height.

~~3206.9~~ 3206.10.3 Dead-end aisles.

~~3206.10~~ 3206.11 Portable fire extinguishers.

3207 SOLID-PILED AND SHELF STORAGE *....no change*

3208 RACK STORAGE

3208.1 General.

3208.2 Fire protection.

3208.2.1 Plastic shelves.

3208.2.2 Racks with solid shelving.

3208.2.2.1 Fire protection.

3208.3 Flue spaces.

3208.3.1 Flue space protection.

3208.4 Column protection.

3208.5 Extra-high-rack storage systems.

3208.5.1 Fire protection.

3209 AUTOMATED STORAGE *....no change*

3210 SPECIALTY STORAGE *....no change*

CHANGE SIGNIFICANCE: This chapter in the code had not changed much since the original printing in the 2000 IFC. Fire testing has occurred and revisions to NFPA 13, "Standard for the Installation of Sprinkler Systems" were made in the meantime. This revision was a review of the entire chapter and provides an update to current engineering practices. The following is a discussion of the major revisions.

Commodity classifications found in the 2015 IFC were not up to date with recent fire testing. The commodity classification criteria was completely reworked in NFPA 13. The revisions in the IFC include a new table for commodity classification. Table 3203.8 lists hundreds of commodities and in many cases addresses the specific packaging materials. The classifications listed are based on the IFC criteria. When a commodity is considered a high-hazard commodity, the table also provides the corresponding NFPA 13 classification since NFPA 13 does not have a high-hazard classification. For example, wood matches in cartons is classified as high-hazard when applying the IFC requirements; it is also listed as Group A unexpanded plastic for the sprinkler design in NFPA 13.

For commodities that are a mixture of Group A plastics and products with other classifications, the percentage of weight and volume of Group A plastic is still evaluated. The single figure in the 2015 IFC has been replaced with two figures in the 2018 edition, along with the inclusion of three equations to assist in the evaluation. This was done to reduce confusion in application of the table. Figure 3203.9(1) addresses the volume of Group A plastic, while Figure 3203.9(2) is assessing the weight of Group

Chapter 32 continues

Chapter 32 continued

A plastics. It is critical to understand that both tables must be utilized and the highest hazard level resulting from either table becomes the commodity classification.

Specific criteria was included in Section 3203.10 to address the use of plastic pallets. Pallets are not included when evaluating the mixed commodity load addressed in the previous paragraph. It is assumed that either wood pallets are used or no pallet, as in solid-pile or shelf storage. But a plastic pallet can add a tremendous fire load to the commodity, unless it is a listed pallet which is considered equivalent to a wood pallet. Depending on the pallet construction, the commodity classification could jump from Class II to Class IV, or from Class IV to High-hazard. The commodity modification based on plastic pallets would affect the sprinkler design, and could also impact the requirements in the IFC.

In Section 3206.3.1, the method for determining the size of the high-piled storage area has been clarified to specifically state that it not only includes aisles within the storage area, but also includes aisle space around the storage. The distance outside of the storage footprint is to be

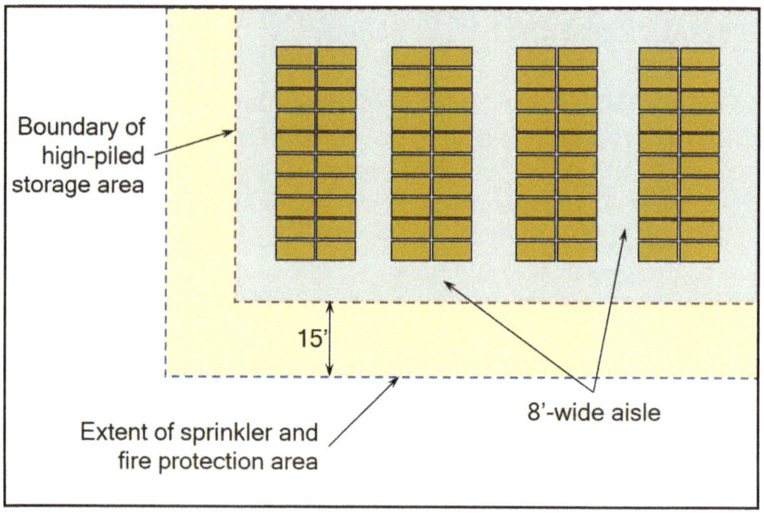

When determining the size of the high-piled combustible storage area, include a perimeter aisle or the distance to a wall, whichever is less.

either the width of an aisle, or the distance to a wall, whichever is less. See the figure below.

In Section 3206.3.2, the requirements for separating high-piled storage areas have been revised. Either all the high-piled storage areas are considered aggregately for determining the requirements in Table 3206.2, or the areas can be considered separately if separated by a 1-hour fire barrier or a distance of 100 feet. The separation of 100 feet does not need to be completely empty, but it cannot contain high-piled combustible storage. It could be a conveyor system, or process area for example. See the figure below.

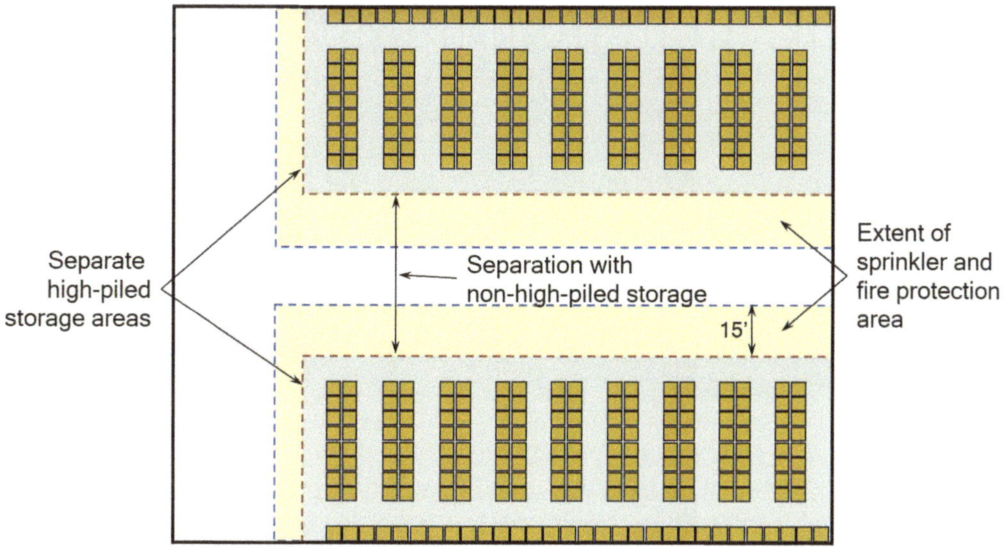

Separating high-piled combustible storage areas by 100 feet allows each area to be considered separately.

Section 3206.7 has been revised for fire department access doors. Access doors are required in exterior walls surrounding the high-piled storage area that face a fire apparatus access road and can now be up to 125 feet apart. This increase is a result of common concrete tilt-up construction using panels of 54 feet wide. This will allow every other panel to be poured the same. An allowance has also been added for narrow buildings, such that access doors are not required if the exterior wall does not face an apparatus road and the opposite wall is less than 150 feet away and is provided with access doors, and all portions of the wall are within 150 feet of a fire department access door.

Table 3208.3 was reformatted to simplify and clarify the requirements for flue spaces. It now clearly indicates it only applies to rack storage arrangements which are protected with an automatic sprinkler system.

3304.5, 3308, 3309.1

Fire Watch During Construction and Demolition

CHANGE TYPE: Modification

CHANGE SUMMARY: Criteria for requiring fire watch has been added to the code along with clarification to the functions and duties of the fire watch personnel.

2018 CODE: **3304.5 Fire watch.** ~~Where required by the fire code official for building demolition, or building construction during working hours that is hazardous in nature, qualified personnel shall be provided to serve as an on-site fire watch. Fire watch personnel shall be provided with not less than one approved means for notification of the fire department and their sole duty shall be to perform constant patrols and watch for the occurrence of fire.~~ Where required by the *fire code official* or the prefire plan established in accordance with Section 3308.3, a fire watch shall be provided for building demolition and building construction that is hazardous in nature, such as temporary heating or hot work.

3304.5.1 Fire watch during construction. Where required by the *fire code official*, a fire watch shall be provided during nonworking hours for new construction that exceeds 40 feet (12 192 mm) in height above the lowest adjacent grade.

3304.5.2 Fire watch personnel. Trained personnel shall be provided to serve as an on-site fire watch. Fire watch personnel shall be provided with not fewer than one *approved* means for notification of the fire department and the sole duty of such personnel shall be to perform constant patrols and watch for the occurrence of fire. The combination of fire watch duties and site security duties is acceptable. Fire watch personnel shall be trained in the use of portable fire extinguishers.

International Code Council®

The fire code official is authorized to require fire watch for this construction project which exceeds 40 feet in height above the lowest adjacent grade.

3304.5.3 Fire watch location and records. The fire watch shall include areas specified by the prefire plan established in accordance with Section 3308.3. The fire watch personnel shall keep a record of all time periods of duty, including a log entry each time the site was patrolled and each time a structure under construction was entered and inspected. The records and log entries shall be made available for review by the *fire code official* upon request.

3308.1 Program development and maintenance. The *owner* or *owner's* authorized agent shall be responsible for the development, implementation and maintenance of a written plan establishing a fire prevention program at the project site applicable throughout all phases of the construction, repair, *alteration* or demolition work. The plan shall address the requirements of this chapter and other applicable portions of this code, the duties of staff, and staff training requirements. The plan shall be made available for review by the *fire code official* upon request.

~~3308.1~~ 3308.2 Program superintendent. The *owner* shall designate a person to be the fire prevention program superintendent who shall be responsible for the fire prevention program and ensure that it is carried out through completion of the project. The fire prevention program superintendent shall have the authority to enforce the provisions of this chapter and other provisions as necessary to secure the intent of this chapter. Where guard service is provided in accordance with NFPA 241, the superintendent shall be responsible for the guard service.

~~3308.3~~ 3308.4 Training. Training of responsible personnel in the use of fire protection equipment shall be the responsibility of the fire prevention program superintendent. Records of training shall be kept and made a part of the written plan for the fire prevention program.

~~3308.4~~ 3308.5 Fire protection devices. The fire prevention program superintendent shall determine that all fire protection equipment is maintained and serviced in accordance with this code. The quantity and type of fire protection equipment shall be *approved*. Fire protection equipment shall be inspected in accordance with the fire prevention program.

3309.1 Emergency telephone. ~~Readily accessible emergency~~ Emergency telephone facilities with *ready access* shall be provided in an *approved* location at the construction site, or an *approved* equivalent means of communication shall be provided. The street address of the construction site and the emergency telephone number of the fire department shall be posted adjacent to the telephone. Alternatively, where an equivalent means of communication has been *approved*, the site address and fire department emergency telephone number shall be posted at the main entrance to the site, in guard shacks and in the construction site office.

CHANGE SIGNIFICANCE: The intent of these revisions is to improve the effectiveness of fire prevention programs on active construction sites. When fires occur during construction and demolition, they can be devastating since many of the fire resistive components of building are not

3304.5, 3308, 3309.1 continues

3304.5, 3308, 3309.1 continued

yet complete and fire protection systems are not active. As a result, NFPA 241, "Standard for Safeguarding Construction, Alteration and Demolition Operations" has undergone several revisions and enhancements and is specifically referenced in these revisions.

Fire watch must be provided when required by the fire code official, or when the prefire plan for the construction operation indicates that fire watch is needed. Potential triggers have been added to assist the fire code official in making this determination. These issues are:

- Construction exceeds 40 feet in height.
- Temporary heating equipment is being used.
- Hot work is planned during the construction.

Once the determination for fire watch has been made, additional criteria has been added regarding the fire watch function. First, the fire watch personnel must be trained. This training must include training on the use of fire extinguishers. Second, the fire watch personnel must maintain a log indicating the time and frequency of their patrol rounds.

The code has traditionally stated that fire watch personnel can have no other duties. The revision in Section 3304.5.2 allows the fire watch personnel to also provide security duties. Many sites already provide a guard service for security, and this revision will allow these functions to be combined with the appropriate additional training.

Section 3308.1 is added with a requirement to develop a written plan. This plan becomes the prefire plan, which may or may not include fire watch as a mitigation factor.

Section 3309.1 is revised to acknowledge the fact that cell phones and radio communication can be acceptable alternatives to hardline phones.

Chapter 38

Higher Education Laboratories

CHANGE TYPE: Addition

CHANGE SUMMARY: A new chapter has been added to the IFC to specifically regulate college and university laboratories. Correlating sections have been added to a new Section 427 in the IBC.

2018 CODE: Due to the extent of the code text, an outline of the chapter is shown below. For the full text, see the 2018 *International Fire Code*.

CHAPTER 38
HIGHER EDUCATION LABORATORIES

3801 GENERAL

3801.1 Scope.

3801.2 Application.

3802 DEFINITIONS

3802.1 Definitions.

3803 GENERAL SAFETY PROVISIONS

3803.1 Scope.

3803.1.1 Chemical safety reviews.

3803.1.2 Chemical handling.

3803.1.3 Warning signage.

International Code Council®

Laboratories at colleges and universities are specifically regulated in Chapter 38.

Chapter 38 continues

Chapter 38 continued

3803.1.4 Maintenance of equipment, machinery and processes.

3803.1.5 Time-sensitive materials.

3803.1.6 Hazardous wastes.

3803.1.7 Automatic fire-extinguishing systems.

3803.2 Hazardous materials storage and use.

3803.2.1 Container size.

3803.2.2 Density.

3804 LABORATORY SUITE CONSTRUCTION

3804.1 General.

3804.1.1 Laboratory suites.

Table 3804.1.1 Design and Number of Laboratory Suites per Floor

3804.1.1.1 Separation from other nonlaboratory areas.

3804.1.1.2 Separation from other laboratory suites.

3804.1.1.3 Floor assembly fire-resistance.

3804.1.1.4 Maximum number.

3804.1.1.5 Means of egress.

3804.1.1.6 Standby or emergency power.

3804.1.1.7 Ventilation.

3804.1.1.8 Liquid-tight floor.

3804.1.1.9 Automatic fire-extinguishing systems.

3804.1.2 Percentage of maximum allowable quantity in each laboratory suite.

3805 NONSPRINKLERED LABORATORIES

3805.1 Scope.

3805.2 Nonsprinklered laboratories.

3805.2.1 Restricted materials storage.

3805.2.2 Restricted materials use.

3805.3 Restricted materials automatic fire detection.

3805.3.1 System supervision and monitoring.

3805.4 Percentage of maximum allowable quantity per control area.

Table 3805.4 Design and Number of Control Areas in Existing Nonsprinklered Laboratories

3806 EXISTING SPRINKLERED LABORATORIES

3806.1 Scope.

3806.2 Hazardous materials storage and use.

3806.2.1 Percentage of maximum allowable quantity per control area.

Table 3806.2.1 Design and Number of Control Areas in Existing Sprinklered Laboratories

SECTION 202
DEFINITIONS

New definition added as follows:

HIGHER EDUCATION LABORATORY. Laboratories in Group B occupancies used for educational purposes above the 12th grade. Storage, use and handling of chemicals in such laboratories shall be limited to purposes related to testing, analysis, teaching, research or developmental activities on a nonproduction basis.

CHANGE SIGNIFICANCE: Higher education academic institutions are quite possibly one of the most important industries to the advancement of technology, science, and medicine. These academic institutions often have chemistry, biology, medical, engineering, and other laboratories where hazardous materials are used. The 2015 IFC does not specifically address teaching and research laboratories, so users must try to apply general hazardous materials provisions, which often are not appropriate for specialized academic laboratory settings.

Several conditions are typically present in academic laboratories that make them unique:

1. In a teaching and research environment, there are often many small laboratories within a building that are using small quantities of hazardous materials in each location. This lower chemical density assists in mitigating the overall risk.
2. Many higher education institutions have a full cadre of faculty and staff with chemical expertise. These "special experts" often include campus fire marshals, industrial hygienists, radiation safety officers, biological safety officers, chemical hygiene officers and environmental health and safety officers who are an integral part of the preparation and review of laboratory safety documentations.
3. A typical university science building will house laboratories, office space, storerooms, classrooms, and lecture halls. The current limits on hazardous materials are so restrictive on upper floors that many universities are forced to locate classrooms and lecture halls on the upper floors so that they can take full advantage of the hazardous materials quantities allowed on the lower floors. This results in moving large numbers of students through hallways, past laboratories to get to the upper floors. They will also have to exit back down past the laboratories in the event of an emergency.

This chapter addresses higher education laboratories and provides a balanced approach to mitigate the hazards, including administrative and emergency planning requirements and construction and fire safety provisions. Specifically, the chapter addresses three primary needs: 1) increasing general laboratory safety requirements, 2) increasing the maximum allowable quantity in large or multistory laboratories, or laboratories located in multistory buildings, and 3) allowing very small quantities of currently prohibited hazardous materials in nonsprinklered laboratory buildings.

Chapter 38 continues

Chapter 38 continued

This chapter introduces the concept of laboratory suites. Laboratory suites are a design alternative and can be used with or without the traditional control area concepts. The "laboratory suite" concept provides an option to allow more flexibility in hazardous materials use, in exchange for additional administrative and construction safeguards, while still remaining a Group B occupancy. Laboratory suites are separated from each other with fire-resistance-rated construction, and are also required to be separated from the nonlaboratory areas of the building. The new Table 3804.1.1 limits the number of laboratory suites per floor and provides the allowable percentage of the maximum allowable quantities. These percentages are greater than those found in Section 5003.8.3.2, but are only allowed when the additional safeguards are provided.

Additional safeguards are added by limiting the size of containers to a maximum of 5.3 gallons for liquids, 50 pounds for solids, 500 cubic feet for physical hazard gases, and 100 cubic feet for health hazard gases.

Section 3805 addresses the use of hazardous materials in existing laboratories and is applicable to an existing building with an existing laboratory. However, when a new laboratory is added to an existing building, the building must comply with the requirements for new construction. This is specifically addressed in Section 3803.1.8 which requires a fire sprinkler system in new laboratories in new or existing buildings if the lab will be using the increased quantities allowed in this chapter.

Existing nonsprinklered laboratories can continue to comply with the quantity limitations in Chapter 50 based on the control area concept, or if they desire to use the quantities allowed in Chapter 38, they must install either a fire sprinkler system or a fire detection system. If they opt for the fire detection system, the quantities are reduced to 25 percent of the allowable quantity permitted in a sprinklered laboratory.

Chapter 39

Processing and Extraction Facilities

CHANGE TYPE: Addition

CHANGE SUMMARY: A new chapter has been added to the IFC to specifically regulate the process of extracting oils from plant material.

2018 CODE: Due to the extent of the code text, an outline of the chapter is shown below. For the complete text, see the 2018 *International Fire Code.*

International Code Council®

The extraction process can use flammable gas, flammable liquids, or combustible liquids as the media for oil extraction.

CHAPTER 39
PROCESSING AND EXTRACTION FACILITIES

3901 GENERAL

- **3901.1 Scope.**
- **3901.2 Existing buildings or facilities.**
- **3901.3 Permits.**

3902 DEFINITIONS

- **3902.1 Definitions.**

3903 PROCESSING AND EXTRACTION

- **3903.1 Construction.**
- **3903.2 Prohibited occupancies.**
- **3903.3 Location.**
- **3903.4 Post-process purification and winterization.**
  - **3903.4.1 Industrial ovens.**
- **3903.5 Use of flammable and combustible liquids.**
- **3903.6 Liquefied petroleum gas.**

3904 SYSTEMS AND EQUIPMENT

- **3904.1 General requirements.**
- **3904.2 Systems and equipment.**
- **3904.3 Technical report.**
  - **3904.3.1 Report content.**
- **3904.4 Site inspection.**

3905 SAFETY SYSTEMS

- **3905.1 Gas detection.**
  - **3905.1.1 System design.**
  - **3905.1.2 Gas detection system components.**
  - **3905.1.3 Operation.**

Chapter 39 continues

Chapter 39 continued

3905.1.4 Failure of the gas detection system.

3905.1.5 Interlocks.

3905.2 Emergency shutoff.

CHANGE SIGNIFICANCE: This new chapter establishes regulations for the extraction of oil from plant materials using solvents as the medium. This chapter was developed in response to the rapid growing industry of cannabis oil extraction. However, the solvent extraction processes are the same whether it is cannabis or other plant materials, so the provisions are not limited to cannabis oil.

Oil extraction activities pose serious risks. The solvents utilized in the extraction process are flammable or combustible liquids or flammable gases, which create inherent hazards in the extraction process. Oil extraction using solvents can involve processes that pose serious risks to public health, safety, and welfare, as illustrated by the 2013 explosions, fires, and fatalities in Bellevue and Spokane, WA.

This chapter requires operational and construction permits for oil extraction. As part of the approval process, the location and construction of the building is evaluated. This elevated hazard is not allowed in Group A, E, I, or R occupancies.

Definitions for "Miscella" and "Desolventizing" are added to the code. The miscella is the material removed during the extraction process and contains the oil and residual solvents. The desolventizing process is the removal of the solvents from miscella to obtain the final oil product. Both of the materials and the associated processes present significant hazards which must be evaluated.

Section 3904 regulates the equipment and systems, and it requires that all equipment for the oil extraction process be listed or approved. There is little, if any, listed equipment for these types of processes. Therefore, where nonlisted equipment is utilized, the systems and process must be reviewed by a registered design professional to evaluate the safety features in the design. The design professional is to provide a technical report evaluating the safety and operation of the process. Section 3904.4 states the fire code official can require the person preparing the technical report to inspect the site to ensure that the installation complies with the design documents and the technical analysis.

PART 5

Hazardous Materials

Chapters 50 through 67

- **Chapter 50** Hazardous Materials—General Provisions
- **Chapter 51** Aerosols
- **Chapter 52** Reserved
- **Chapter 53** Compressed Gases
- **Chapter 54** Corrosive Materials No changes addressed
- **Chapter 55** Cryogenic Fluids No changes addressed
- **Chapter 56** Explosives and Fireworks No changes addressed
- **Chapter 57** Flammable and Combustible Liquids
- **Chapter 58** Flammable Gases and Flammable Cryogenic Fluids No changes addressed
- **Chapter 59** Flammable Solids No changes addressed
- **Chapter 60** Highly Toxic and Toxic Materials No changes addressed
- **Chapter 61** Liquefied Petroleum Gases
- **Chapter 62** Organic Peroxides No changes addressed
- **Chapter 63** Oxidizers, Oxidizing Gases and Oxidizing Cryogenic Fluids No changes addressed
- **Chapter 64** Pyrophoric Materials No changes addressed
- **Chapter 65** Pyroxylin (Cellulose Nitrate) Plastics No changes addressed
- **Chapter 66** Unstable (Reactive) Materials No changes addressed
- **Chapter 67** Water-Reactive Solids and Liquids No changes addressed
- **Chapters 68 through 79** (Reserved)

Chapters 50 through 67 contain requirements for the safe storage, use, handling and dispensing of hazardous materials. The Chemical Abstract Service, a part of the American Chemical Council, has identified over 35 million organic and inorganic substances and over 60 million sequences or mixtures. The *International Fire Code* (IFC) does not regulate all of these chemicals. Instead, the IFC focuses on chemicals or chemical mixtures that pose physical or health hazards. Physical hazard materials can detonate, deflagrate, burn or accelerate burning. Health hazard materials can incapacitate, injure or cause death after only a single exposure. Chapters 68 through 79 are reserved and serve as place holders for future code provisions when new hazardous materials are identified as requiring regulation by the IFC. ■

TABLE 5003.1.1(1)

Consumer Fireworks

5003.1.1(1), 5003.11.1, 6303.1.1.2

Maximum Allowable Quantity for Class 3 Oxidizers

5003.8.3.4

Construction of Control Areas

5005.1.12

Protection of Hazardous Materials Piping Systems

5103.2, 5104.1.2

Aerosol Products in Plastic Containers

5103.2.2, 5104.2.2, 5104.3.3, 5104.8, 5106.2.2

Aerosol Cooking Spray Products

5306.1, 5306.2

Medical Gas Storage

5307.1, 5307.3

Liquid Carbon Dioxide Systems for Beverage Dispensing

5307.4

Carbon Dioxide Enrichment Systems

5707

Mobile Fueling Operations

6104.3

Location of LP-gas Containers

CHANGE TYPE: Modification

CHANGE SUMMARY: Addresses the explosive nature of Division 1.4G explosives and removes the 100 percent increase in quantity for sprinklers where these items are stored.

Table 5003.1.1(1)

Consumer Fireworks

2018 CODE:

TABLE 5003.1.1(1) (*PARTIAL*) Maximum Allowable Quantity per Control Area of Hazardous Materials Posing a Physical Hazard[a, j, m, n, p]

MATERIAL	CLASS	GROUP WHEN THE MAXIMUM ALLOWABLE QUANTITY IS EXCEEDED	STORAGE[b]			USE-CLOSED SYSTEMS[b]			USE-OPEN SYSTEMS[b]	
			Solid pounds (cubic feet)	Liquid gallons (pounds)	Gas (cubic feet at NTP)	Solid pounds (cubic feet)	Liquid gallons (pounds)	Gas (cubic feet at NTP)	Solid pounds (cubic feet)	Liquid gallons (pounds)
~~Consumer fireworks~~	~~1.4G~~	~~H-3~~	~~125[e, l]~~	~~NA~~	~~NA~~	~~NA~~	~~NA~~	~~NA~~	~~NA~~	~~NA~~
Explosives	Division 1.1	H-1	1[e, g]	(1)[e, g]		0.25[g]	(0.25)[g]		0.25[g]	(0.25)[g]
	Division 1.2	H-1	1[e, g]	(1)[e, g]		0.25[g]	(0.25)[g]		0.25[g]	(0.25)[g]
	Division 1.3	H-1 or H-2	5[e, g]	(5)[e, g]		1[g]	(1)[g]		1[g]	(1)[g]
	Division 1.4	H-3	50[e, g]	(50)[e, g]	NA	50[g]	(50)[g]	NA	NA	NA
	Division 1.4G	H-3	125 [~~d~~, e, l]	NA		NA	NA		NA	NA
	Division 1.5	H-1	1[e, g]	(1)[e, g]		0.25[g]	(0.25)[g]		0.25[g]	(0.25)[g]
	Division 1.6	H-1	1[e, g]	NA		NA	NA		NA	NA

SECTION 202 DEFINITIONS

EXPLOSIVE. A chemical compound, mixture or device, the primary or common purpose of which is to function by explosion. The term includes, but is not limited to, dynamite, black powder, pellet powder, initiating explosives, detonators, safety fuses, squibs, detonating cord, igniter cord~~;~~ and igniters. ~~and display fireworks, 1.3G.~~

The term "Explosive" includes any material determined to be within the scope of USC Title 18: Chapter 40 and also includes any material classified as an explosive ~~other than consumer fireworks, 1.4G~~ by the hazardous materials regulations of DOTn 49 CFR Parts 100-185.

High explosive. *No change to text...*

Low explosive. *Explosive material* that will burn or deflagrate when ignited. It is characterized by a rate of reaction that is less than the speed of sound. Examples of low *explosives* include, but are not limited to, black powder, safety fuse, igniters, igniter cord, fuse lighters, fireworks, ~~1.3G~~ and propellants, 1.3C.

Mass-detonating explosives. *No change to text...*

Table 5003.1.1(1) continues

Table 5003.1.1(1) continued

UN/DOTn Class 1 explosives. *No change to text…*

Division 1.1. *No change to text…*

Division 1.2. *No change to text…*

Division 1.3. *No change to text…*

Division 1.4. *No change to text…*

Division 1.5. *No change to text…*

Division 1.6. *No change to text…*

FIREWORKS. Any composition or device for the purpose of producing a visible or an audible effect for entertainment purposes by combustion, *deflagration* or *detonation* that meets the definition of 1.3G fireworks or 1.4G fireworks.

Fireworks, 1.3G. Large fireworks devices, which are *explosive materials*, intended for use in fireworks displays and designed to produce audible or visible effects by combustion, *deflagration* or *detonation*. Such 1.3G fireworks include, but are not limited to, firecrackers containing more than 130 milligrams (2 grains) of explosive composition, aerial shells containing more than 40 grams of pyrotechnic composition and other display pieces which exceed the limits for classification as 1.4G fireworks. Such 1.3G fireworks are also described as Fireworks, UN 0335 by the DOTn.

Fireworks, 1.4G. Small fireworks devices containing restricted amounts of pyrotechnic composition designed primarily to produce visible or audible effects by combustion~~. Such 1.4G fireworks which comply~~ or deflagration that complies with the construction, chemical composition and labeling regulations of the DOTn for Fireworks, UN

International Code Council®

These fireworks are classified as explosives, 1.4G and would not receive quantity increase based on fire sprinklers in the building.

0336, and the U.S. Consumer Product Safety Commission as set forth in CPSC 16 CFR Parts 1500 and 1507.~~are not explosive materials for the purpose of this code.~~

CHANGE SIGNIFICANCE: These combined changes address a number of hazards associated with Division 1.4G explosives—consumer fireworks. Rather than providing a special exception for these items in the definitions for Explosives, they are consolidated into their own category under the Fireworks definition and Chapter 56 requirements.

Most significant is the change to Table 5003.1.1 – Maximum Allowable Quantity per Control Area of Hazardous Materials Posing a Physical Hazard. Since its inception, it has allowed increases for stored materials when the control area is protected by sprinklers or the materials are in approved storage cabinets. Historically, the quantity increase has been permitted to be 100 percent for each protection method and 200 percent for both. However, explosives and fireworks did not always benefit from the sprinkler increase. Code Change F122-00 added the sprinkler increase to Division 1.4G, consumer fireworks.

Consumer fireworks are removed from the table entirely. This leaves consumer fireworks falling under the classification of explosives, 1.4G. Footnote d is removed from the storage allowance for explosives, 1.4G. This removes the 100-percent increase (from 125 to 250 pounds) in a sprinklered building. Where the storage quantity exceeds 125 pounds per control area, the space must be classified as a Group H, Division 3 occupancy.

These revisions resulted from questioning the logic of the sprinkler increase that went into the code for Division 1.4G without corresponding fire test data to support it.

The Tables of Distances all included an exception that removed 1.4G from the requirements in the tables. That exception is now deleted in the following tables:

1. Table 5601.8.1(3) – Application of Separation Distance (Q-D) Tables—Division 1.4 Explosives

2. Table 5604.3 – Storage Amounts and Magazine Requirements for Explosives, Explosive Materials and Fireworks, 1.3G Maximum Allowable Quantity per Control Area

3. Table 5604.5.2(3) – Table of Distances (Q-D) for Buildings and Magazines Containing Explosives – Division 1.4

5003.1.1(1), 5003.11.1, 6303.1.1.2

Maximum Allowable Quantity for Class 3 Oxidizers

CHANGE TYPE: Modification

CHANGE SUMMARY: The maximum allowable quantity for Class 3 oxidizers is increased by about 10 percent in control areas and Groups M and S occupancies

2018 CODE: **Table 5003.1.1(1) Footnote k.**

k. A maximum quantity of ~~200~~ 220 pounds of solid or ~~20~~ 22 gallons of liquid Class 3 oxidizers is allowed where such materials are necessary for maintenance purposes, operation or sanitation of equipment where the storage containers and the manner of storage are approved.

6303.1.1.2 Class 3 liquid and solid oxidizers. Not more than ~~200~~ 220 pounds (~~91~~ 99 kg) of solid or ~~20~~ 22 gallons (~~76~~ 83 L) of liquid Class 3 oxidizer is allowed in storage and use where such materials are necessary for maintenance purposes or operation of equipment. The oxidizers shall be stored in *approved* containers and in an *approved* manner.

TABLE 5003.11.1 (*PARTIAL*) **Maximum Allowable Quantity per Indoor and Outdoor Control Area in Group M and S Occupancies—Nonflammable Solids, Nonflammable and Noncombustible Liquids**

CONDITION		MAXIMUM ALLOWABLE QUANTITY PER CONTROL AREA	
Material[a]	Class	Solids (pounds)	Liquids (gallons)
B. PHYSICAL-HAZARD MATERIALS—NONFLAMMABLE AND NONCOMBUSTIBLE SOLIDS AND LIQUIDS			
1. Oxidizers[b,c]	4	Not Allowed	Not Allowed
	3	~~1,105[g]~~ 1,350[g]	135 ~~115~~
	2	2,250[h]	225
	1	18,000[i,j]	1,800[i,j]

CHANGE SIGNIFICANCE: The new language recognizes changes that have occurred in Class 3 oxidizer packaging methods to align with international metric (SI) weight and volume capacities.

Class 3 oxidizers can severely increase the burning rate of combustible materials with which they come in contact and will cause sustained and vigorous decomposition if contaminated with a combustible material or if exposed to sufficient heat.

Examples of Class 3 oxidizers include ammonium dichromate, hydrogen peroxide (52 to 91 percent by weight), nitric acid, fuming (concentration greater than 86 percent), perchloric acid solutions (60 to 72 percent by weight), potassium bromate, potassium chlorate, potassium dichloroisocyanurate, sodium chlorate, sodium chlorite (greater than 40 percent by weight) and sodium dichloroisocyanurate.

With the packaging changes, a former 50 pound (22.7 kg) solid oxidizer now weighs 55 pounds (25 kg). Four containers in use (220 pounds or 100 kg) would exceed the maximum allowable quantity per control area (MAQ).

When the hazardous materials MAQ were established for Group M and S occupancies, they were based on the total weight of a pallet load of the specific materials. For example, the total weight of a typical Class 3 solid oxidizer on a pallet was 1,150 pounds (520 kg) so the original MAQ reflected that. Now, normal pallet configuration of eight 55-pound (25kg) pails per row, three high, equals 1,320 pounds (600 kg). Facility operators would have to remove the top row of product from a pallet to comply with the MAQ.

5003.8.3.4

Construction of Control Areas

CHANGE TYPE: Modification

CHANGE SUMMARY: The fire-resistance rating of the floor of a control area in a Type IV building is allowed to be reduced to 1-hour if the building is sprinklered and does not exceed three stories in height.

2018 CODE: **5003.8.3.4 Fire-resistance rating requirements.** The required *fire-resistance rating* for *fire barriers* shall be in accordance with Table 5003.8.3.2. The floor assembly of the *control area* and the construction supporting the floor of the *control area* shall have a *fire-resistance rating* of not less than 2 hours.

> **Exception:** The floor assembly of the *control area* and the construction supporting the floor of the *control area* is allowed to be 1-hour *fire-resistance rated* in buildings of Types IIA, IIIA, IV and VA construction, provided that both of the following conditions exist:
>
> 1. The building is equipped throughout with an *automatic sprinkler system* in accordance with Section 903.3.1.1.
> 2. The building is three stories or less above grade plane.

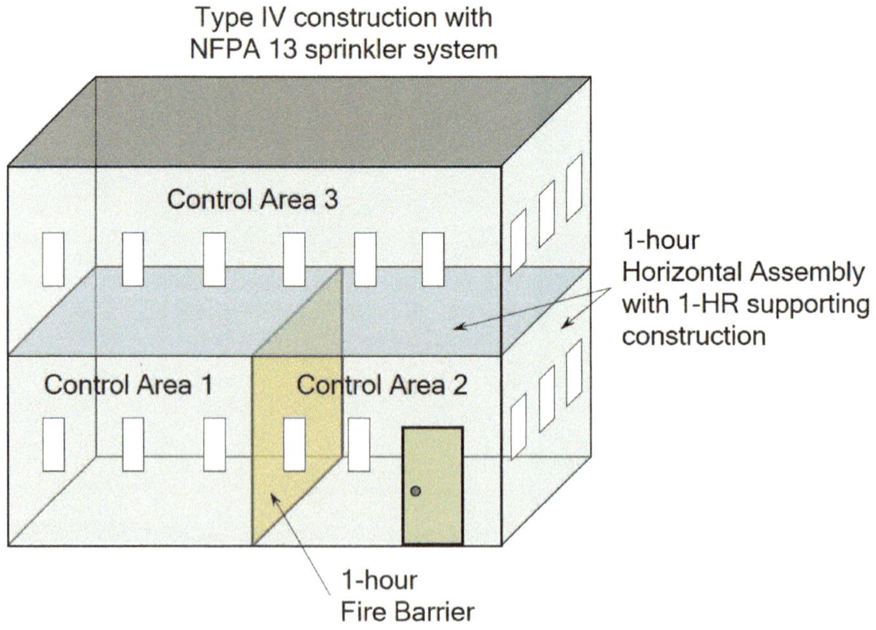

The floor of Control Area 3 in this building of Type IV construction can be 1-hour construction when the building has an automatic sprinkler system.

CHANGE SIGNIFICANCE: Recent interest in "green" building construction materials has prompted a code review to see where changes might be made to accommodate it. This change recognizes that the *International Buliding Code* (IBC) hourly fire barrier assembly requirements for floors in Type IV construction are the same as those required in Types IIIA and VA buildings.

IBC Section 2304.11.3 provides prescriptive construction details for floor assemblies of heavy timber construction, including sawn wood, glued-laminated plank floors or cross-laminated timber (CLT) floors. Concealed spaces are not permitted in any of the assemblies.

There are two installation options for sawn or glued-laminated plank floors. In one, an underlying layer of splined or tongue-and-grooved planks must be at least 3 inches thick. The second requires planks at least 4 inches wide to be set on edge (creating a minimum 4-inch thick floor).

Both options must be covered with 1-inch thick flooring, $^{15}/_{32}$-inch structural wood panels, or $^{1}/_{2}$-inch particleboard. The floors must be laid so no continuous joint lines occur except at support points (underlying beams and columns). This arrangement minimizes the likelihood that hot gases or flames will penetrate the assembly.

CLT floors must be at least 4 inches thick and continuous from support to support. The CLT members must be nailed or screwed to one another.

This method of Type IV, or heavy timber, construction creates an inherent fire resistance just from the sheer mass of the wood materials. As these wood assemblies are exposed to flames, they develop a char layer that aids in the insulation of the remaining wood from further flame impingement. This allows an equivalency to 1-hour fire-resistance-rated assemblies that is required for Types IIA, IIIA and VA construction.

5005.1.12

Protection of Hazardous Materials Piping Systems

International Code Council®

Pneumatically operated shutoff valve on a gas cylinder.

CHANGE TYPE: Modification

CHANGE SUMMARY: Requirements for leak detection and emergency shutoff for high hazard gases and liquids only apply when the maximum allowable quantity per control area is exceeded.

2018 CODE: **5003.2.2.1 Design and construction.** Piping, tubing, valves, fittings and related components used for hazardous materials shall be in accordance with the following:

1. Piping, tubing, valves, fittings and related components shall be designed and fabricated from materials that are compatible with the material to be contained and shall be of adequate strength and durability to withstand the pressure, structural and seismic stress and exposure to which they are subject.
2. Piping and tubing shall be identified in accordance with ASME A13.1 to indicate the material conveyed.
3. ~~Readily accessible~~ Manual valves or automatic remotely activated fail-safe emergency shutoff valves shall be installed on supply piping and tubing and provided with *ready access* at the following locations:
    3.1. The point of use.
    3.2. The tank, cylinder or bulk source.
4. Manual emergency shutoff valves and controls for remotely activated emergency shutoff valves shall be identified and the location shall ~~be~~ have access clearly visible~~, accessible~~ and indicated by means of a sign.
5. Backflow prevention or check valves shall be provided where the backflow of hazardous materials could create a hazardous condition or cause the unauthorized discharge of hazardous materials.

~~6. Where gases or liquids having a hazard ranking of:~~

~~Health Class 3 or 4~~

~~Flammability Class 4~~

~~Instability Class 3 or 4~~

~~in accordance with NFPA 704 are carried in pressurized piping above 15 pounds per square inch gauge (psig) (103 kPa), an approved means of leak detection and emergency shutoff or excess flow control shall be provided. Where the piping originates from within a hazardous material storage room or area, the excess flow control shall be located within the storage room or area. Where the piping originates from a bulk source, the excess flow control shall be located as close to the bulk source as practical.~~

Exceptions:

1. Piping for inlet connections designed to prevent backflow.
2. Piping for pressure relief devices.

5005.1.12 Emergency Isolation. Where gases or liquids having a hazard ranking of Health Class 3 or 4, Flammability Class 4 or Instability Class 3 or 4 in accordance with NFPA 704 are carried in pressurized piping above 15 pounds per square inch gauge (psig) (103 kPa), an *approved* means of leak detection and emergency shutoff or excess flow control shall be provided. Where the piping originates from within a hazardous material storage room or area, the excess flow control shall be located within the storage room or area. Where the piping originates from a bulk source, the excess flow control shall be located as close to the bulk source as practical.

Exceptions:

1. Piping for inlet connections designed to prevent backflow.
2. Piping for pressure relief devices.

CHANGE SIGNIFICANCE: Section 5003.2.2.1, Item 6 is relocated to 5005.1.12. While this may not seem like much of a change, because the text is unchanged, it actually has a significant impact. In the 2015 IFC where this section is found in Section 5003, it applies to any quantity of these hazardous materials. In other words, quantities above and below the maximum allowable quantity per control area (MAQ) are required to comply. In the 2018 IFC, these provisions are located in Section 5005. Section 5005 only applies when the MAQ is exceeded for use, dispensing or handling.

This section requires an approved means of leak detection and either an emergency shutoff or excess flow control for the most hazardous classes of gases and liquids when they are pressurized to over 15 psig. Previously, this requirement applied to any amount. Now, this requirement will apply when the MAQ is exceeded. Therefore, the facility must be a Group H occupancy.

Based on IFC Appendix F "Hazard Ranking", these requirements apply to:

NFPA 704 CLASSIFICATION	IFC/IBC MATERIAL CLASSIFICATION
Health Class 4	• Highly toxic gas or liquid
Health Class 3	• Corrosive gas or liquid • Cryogenic fluid • Toxic gas or liquid
Flammability Class 4	• Flammable cryogenic fluid • Flammable gas • Class IA flammable liquid • Class I organic peroxide • Pyrophoric gas
Instability Class 4	• Unclassified detonable organic peroxide • Class 4 detonable unstable reactive • Class 3 detonable unstable reactive
Instability Class 3	• Class I organic peroxide • Class II organic peroxide • Class 3 water reactive

The material list above is just a general application, there can be other materials that will fall into these categories. Each material would need to be evaluated individually based on its characteristics.

5103.2, 5104.1.2

Aerosol Products in Plastic Containers

CHANGE TYPE: Modification

CHANGE SUMMARY: Limitations on aerosol products in plastic containers is revised and the use of Plastic Aerosol X products is prohibited in higher life hazard occupancies.

2018 CODE: **5103.2 Identification.** Cartons or outer packaging shall be identified on not ~~less~~ fewer than one exterior side with the classification level of the aerosol products contained within the carton. ~~as follows~~:

5103.2.1 Aerosol products. Cartons or outer packaging containing aerosol products in metal containers or glass and plastic containers 4 fluid ounces (118 ml) or less shall be clearly marked as follows:

LEVEL ______ AEROSOLS

5103.2.3 Plastic aerosol products. Cartons or outer packaging containing aerosol products in plastic containers greater than 4 fluid ounces (118 ml) shall be clearly marked as follows:

PLASTIC AEROSOL 1 (or X)

5104.1.2 Plastic aerosol X products. Plastic aerosol X products are those products, in containers larger than 4 fluid ounces (118 ml), that do not meet the criteria provided in Section 5104.1.1.

5104.1.2.1 Storage, use or handling. The storage, use or handling of plastic aerosol X products shall be prohibited.

SECTION 202
DEFINITIONS

AEROSOL CONTAINER. A metal can or plastic container up to a maximum size of 33.8 fluid ounces (1000 ml) or a glass ~~or plastic~~ bottle up to a maximum size of 4 fluid ounces (118 ml) designed and intended to dispense an aerosol.

CHANGE SIGNIFICANCE: Aerosol products are available for most industries and provide an easy, quick way to apply, or dispense, a product. The aerosol industry provides aerosol products in traditional metal containers, and plastic or glass containers are also becoming common.

The definition of an aerosol container is revised to specify that a metal or plastic container can be up to one liter in size, while glass containers are limited to 4 fluid ounces.

Packaging containing plastic aerosol containers will be marked in one of three ways:

- Level 1 (or 2, or 3) Aerosol
    - o Section 5103.2.1

- o Container size: plastic containers not greater than 4 fluid ounces
- o Product: could be Level 1, 2, or 3 aerosol products

- Plastic Aerosol 1
  - o Sections 5103.2.3 and 5104.1.1
  - o Container size: plastic containers greater than 4 fluid ounces, but less than 33.8 fluid ounces
  - o Product: must meet the limited flammability criteria in Section 5104.1.1
- Plastic Aerosol X
  - o Section 5104.1.2
  - o Container size: plastic containers greater than 4 fluid ounces, but less than 33.8 fluid ounces
  - o Product: does NOT meet the limited flammability criteria in Section 5104.1.1

Section 5104.1.1 allows aerosol products in plastic containers up to one liter in size provided the products comply with the following criteria:

1. The base product has no fire point and the propellant is nonflammable.
2. The base product does not sustain combustion and the propellant is nonflammable.
3. The base product contains no more than 20 percent by volume (15.8 percent by weight) of ethanol, isopropyl alcohol or a combination thereof in an aqueous mix, and the propellant is nonflammable.
4. The base product contains no more than 4 percent by weight of an emulsified flammable liquefied gas propellant within an aqueous base. The propellant must remain emulsified for the life of the product, or the propellant is nonflammable.

Based on the criteria in Section 5104.1.1, the aerosol products allowed in plastic containers are low to moderate flammability and treated as Class III commodities in accordance with NFPA 13. When the aerosol product exceeds the flammability criteria, it is labeled as Plastic Aerosol X and Section 5104.1.2.1 prohibits its storage, use or handling in Groups A, B, E, F, I, and R occupancies.

See also Significant Change to Section 5104.8 for aerosol cooking spray products.

5103.2.2, 5104.2.2 5104.3.3, 5104.8 5106.2.2

Aerosol Cooking Spray Products

International Code Council®

Aerosol cooking spray product with label on the box.

CHANGE TYPE: Addition

CHANGE SUMMARY: Specific fire protection requirements are added to address aerosol cooking spray products.

2018 CODE: **5103.2.2 Aerosol cooking spray products.** Cartons or outer packaging containing aerosol cooking spray products in metal containers shall be clearly marked as follows:

AEROSOL COOKING SPRAY

5104.2.2 Aerosol cooking spray products. Storage of aerosol cooking spray products in A, B, E, F and R occupancies shall not be more than 1,000 pounds (454 kg) net weight.

5104.3.3 Aerosol cooking spray products. Solid pile, palletized or rack storage of aerosol cooking spray products in a general purpose warehouse shall not be more than 2,500 pounds (1135 kg) net weight, unless protected in accordance with NFPA 30B.

5104.8 Storage of aerosol cooking spray products. *Aerosol cooking spray products* shall be permitted to be stored in a general-purpose warehouse.

5104.8.1 Mixed storage. Where *aerosol cooking spray products* are mixed with other higher-hazard aerosol products, the provided isolation, storage height restrictions and protection shall be based on the highest-hazard aerosol product present.

5104.8.2 Storage conditions. The storage and handling of *aerosol cooking spray products* shall comply with this chapter and NFPA 30B.

5106.2.2 Aerosol cooking spray storage and fire protection. The storage and handling of *aerosol cooking spray products* shall comply with this chapter and NFPA 30B.

SECTION 202
DEFINITIONS

AEROSOL COOKING SPRAY PRODUCTS. Aerosol cooking spray products are those aerosol products designed to deliver a vegetable oil or a solid or nonflammable liquid to reduce sticking on cooking and baking surfaces, or to be applied to food, or both. These products have a chemical heat of combustion that is greater than 8600 Btu/lb (20 kJ/g) and contain no more than 18 percent by weight of flammable propellant.

CHANGE SIGNIFICANCE: Classification of aerosol products was developed by Factory Mutual (now FM Global) in the early 1990s. This was based on a large number of fire tests on palletized aerosol products, a test program obtaining data on various formulas and an array of steel and aluminum container sizes.

Based on all initial testing, it was determined that materials with a flash point over 500°F could be ignored regarding their contribution to an aerosol product fire. However, some aerosol products are designed to deliver a vegetable oil, or like substance, to reduce sticking on cooking or baking surfaces. These products may have very high levels of oils with flash points over 500°F. A fire test of this type of product at a much later date showed that it could not be protected as a Class III commodity. The exemption for these materials with a flash point over 500°F was removed from NFPA 30B.

Subsequently, Factory Mutual conducted large-scale fire tests on these cooking spray products. As a result of these tests, cooking oil products with a chemical heat of combustion equal to or less than 8,600 Btu/lb are still classified, and the carton is marked, as Level 1 Aerosol. However, some cooking oils have a heat of combustion in excess of 8,600 Btu/lb. Therefore, a new definition is added to specify that the cooking oil classification is limited to products with a chemical heat of combustion greater than 8,600 Btu/lb and the flammable propellant does not exceed 18 percent by weight of the product. New fire protection schemes were developed and are found in Tables 6.3.2.3(a) and 6.3.2.3(b) of NFPA 30B. The sprinkler protection design is greater than that required for a Class III commodity, but less than that for Level 2 aerosol products.

Section 5103.2.2 requires that these aerosol products are labeled as AEROSOL COOKING SPRAY. Other aerosol products would be labeled as AEROSOL LEVEL 1 (or 2, or 3) as appropriate.

Aerosol cooking spray products are limited to a maximum quantity of 1,000 pounds in Groups A, B, E, F, and R occupancies. Storage in general-purpose warehouses is limited to 2,500 pounds. Both of these limitations are the same as allowed for Level 2 aerosol products. However, quantities can be increased where the fire protection scheme is in accordance with NFPA 30B, *Code for the Manufacture and Storage of Aerosol Products.*

See also Significant Change to Section 5103.2 for plastic aerosol containers.

5306.1, 5306.2

Medical Gas Storage

CHANGE TYPE: Modification

CHANGE SUMMARY: Requirements for construction and ventilation of interior medical gas rooms and gas cabinets are revised.

2018 CODE: 5306.1 General. Medical gases at health care-related facilities intended for patient ~~care, inhalation or sedation including, but not limited to, analgesia systems for dentistry, podiatry,~~ or veterinary care ~~and similar uses~~ shall comply with Sections 5306.2 through 5306.5 in addition to other requirements of this chapter and Section 427 of the *International Building Code*.

5306.2 Interior supply location. Medical gases shall be ~~stored~~ located in areas dedicated to the storage of such gases without other storage or uses. Where containers of medical gases in quantities greater than the permit amount are located inside buildings, they shall be in a 1-hour exterior room, a 1-hour interior room or a gas cabinet in accordance with Section 5306.2.1, 5306.2.2 or 5306.2.3, respectively. Rooms or areas where medical gases are stored or used in quantities exceeding the *maximum allowable quantity per control area* as set forth in Section 5003.1 shall be in accordance with the *International Building Code* for high-hazard Group H occupancies.

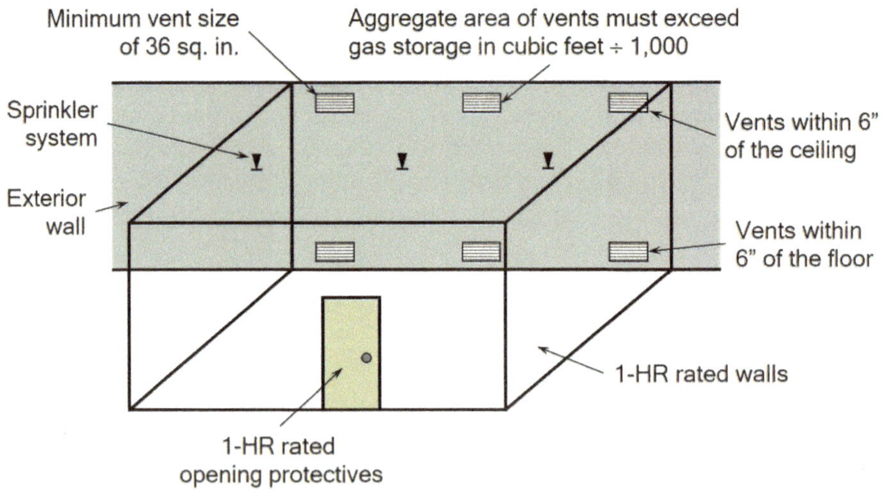

Medical gases can be stored in an exterior gas storage room.

5306.2.1 One-hour exterior rooms. A 1-hour exterior room shall be a room or enclosure separated from the remainder of the building by *fire barriers* constructed in accordance with Section 707 of the *International Building Code* or horizontal assemblies constructed in accordance with Section 711 of the *International Building Code*, or both, with a *fire-resistance rating* of not less than 1 hour. Openings between the room or enclosure and interior spaces shall be self-closing smoke- and draft-control assemblies having a *fire protection rating* of not less than 1 hour. Rooms shall have not less than one exterior wall that is provided with not less than two ~~nonclosable louvered~~ vents. Each vent shall have a minimum free opening area of ~~24~~ 36 square inches (~~155~~ 232 cm^2) for each 1,000 cubic feet (28 m^3) at *normal temperature and pressure* (NTP) of gas stored in the room and shall be not less than 72 square inches (465 cm^2) in aggregate free opening area. One vent shall be within 6 inches (152 mm) of the floor and one shall be within 6 inches (152 mm) of the ceiling. Rooms shall be provided with not less than one automatic sprinkler to provide container cooling in case of fire.

5306.2.2 One-hour interior room. Where an exterior wall cannot be provided for the room, a 1-hour interior room shall be a room or enclosure separated from the remainder of the building by *fire barriers* constructed in accordance with Section 707 of the *International Building Code* or horizontal assemblies constructed in accordance with Section 711 of the *International Building Code*, or both, with a *fire-resistance rating* of not less than 1 hour. Openings between the room or enclosure and interior spaces shall be self-closing smoke- and draft-control assemblies having a *fire protection rating* of not less than 1 hour. An *automatic* ~~sprinklers~~ *sprinkler system* shall be installed within the room. The room shall be exhausted through a duct to the exterior. Supply and exhaust ducts shall be enclosed in a 1-hour-rated shaft enclosure from the room to the exterior. *Approved* mechanical ventilation shall comply with the *International Mechanical Code* and be provided at a minimum rate of 1 cfm per square foot (0.00508 $m^3/(s \bullet m^2)$ of the area of the room.

5306.2.3 Gas cabinets. Gas cabinets shall be constructed in accordance with Section 5003.8.6 and shall comply with the following:

1. Exhausted to the exterior through dedicated exhaust duct system installed in accordance with Chapter 5 of the *International Mechanical Code.*

~~1.~~2. Supply and exhaust ducts shall be enclosed in a 1-hour *fire-resistance-rated* shaft enclosure from the cabinet to the exterior. The average velocity of ventilation at the face of access ports or windows shall be not less than 200 feet per minute (1.02 m/s) with not less than 150 feet per minute (0.76 m/s) at any point of the access port or window.

~~2. They shall be connected to an exhaust system.~~

3. ~~They shall be internally sprinklered.~~ Provided with an *automatic sprinkler system* internal to the cabinet.

CHANGE SIGNIFICANCE: The revision in Section 5306.1 eliminates the laundry list of facilities where medical gas systems are regulated and replaces this list with the terms "patient or veterinary care." This occurred because ambulatory care facilities were not listed, but rather than add them to the list, the general requirements for patient or veterinary care remain. The terms "patient or veterinary care" cover all of the intended uses listed in the 2015 IFC, including ambulatory care facilities.

Construction of medical gas rooms is now regulated in the new Section 427 in the *International Building Code*. The requirements in the IBC are consistent with those in the IFC.

Medical gases are to be stored in either a 1-hour exterior room, a 1-hour interior room, or a gas cabinet when the quantity exceeds the

5306.1, 5306.2 continues

5306.1, 5306.2 continued

permit amount. These rooms could be classified as part of the main occupancy, since the permit amount is less than the maximum allowable quantity per control area. However, when the quantity exceeds the maximum allowable quantity per control area, the storage room is then classified as a Group H occupancy.

When applying these provisions, Section 5306.2.2 states that when an exterior room cannot be provided, then an interior room is acceptable. Both the interior room and the exterior room are required to be separated from the remainder of the building by 1-hour construction, and both require ventilation. Ventilation in the exterior room can be provided with vents located in the exterior wall, but the interior room must be mechanically ventilated.

The size of ventilation openings for the exterior room has been changed from 24 square inches to 36 square inches for each 1,000 cubic feet of gas in the medical gas room.

Also, correlation has been made between the use of gas cabinets and interior gas rooms. The 2015 IFC only required the supply and exhaust ducts of an interior gas room to be located within a 1-hour rated enclosure. When gas cabinets are used as the method of protection for medical gases, the supply and exhaust ducts for the gas cabinets are now required to be within a 1-hour rated enclosure.

5307.1, 5307.3

Liquid Carbon Dioxide Systems for Beverage Dispensing

CHANGE TYPE: Modification

CHANGE SUMMARY: Requirements for liquefied CO_2 in beverage dispensing applications have been correlated with requirements for gas detection systems.

2018 CODE: **5307.1 General.** *Compressed gases* in storage or use not regulated by the material-specific provisions of Chapters 6, 54, 55, and 60 through 67, including asphyxiant, irritant and radioactive gases, shall comply with this section in addition to other requirements of this chapter.

~~**5307.2 Permits.** Permits shall be required as set forth in Section 105.6.~~

~~**5307.3 Equipment.** The storage, use, and handling of liquid carbon dioxide shall be in accordance with Chapter 53 and the applicable requirements of NFPA 55, Chapter 13. Insulated liquid carbon dioxide systems shall have pressure relief devices vented in accordance with NFPA 55.~~

~~**5307.4 Protection from damage.** Carbon dioxide systems shall be installed so the storage tanks, cylinders, piping and fittings are protected from damage by occupants or equipment during normal facility operations.~~

~~**5307.1**~~ **5307.3** ~~**General**~~ **Insulated liquid carbon dioxide systems used in beverage dispensing applications.** ~~Carbon~~ Insulated liquid carbon dioxide systems with more than 100 pounds (45.4 kg) of carbon dioxide used in beverage dispensing applications shall comply with ~~Sections 5307.2 through 5307.5.2~~ Section 5307.3.1.

~~**5307.5**~~ **5307.3.1** ~~**Required protection**~~ **Ventilation.** Where insulated liquid carbon dioxide storage tanks, cylinders, piping and equipment are located indoors, rooms or areas containing ~~carbon dioxide~~ storage tanks, cylinders, piping and ~~fittings~~ equipment, and other areas where a leak of carbon dioxide ~~can collect~~ is expected to accumulate, shall be provided with ~~either~~ mechanical ventilation in accordance with Section ~~5307.5.1 or an emergency alarm system in accordance with Section 5307.5.2~~ 5004.3 and designed to maintain the room containing carbon dioxide at a negative pressure in relation to the surrounding area.

> **Exception:** A gas detection system complying with Section 5307.3.2 shall be permitted in lieu of mechanical ventilation.

~~**5307.5.1 Ventilation.** Mechanical ventilation shall be in accordance with the International Mechanical Code and shall comply with all of the following:~~

~~1. Mechanical ventilation in the room or area shall be at a rate of not less than 1 cubic foot per minute per square foot [0.00508 m³/(s • m²)].~~

5307.1, 5307.3 continues

International Code Council®

Insulated liquid carbon dioxide systems exceeding 100 pounds must be provided with either mechanical ventilation or a gas detection system.

5307.1, 5307.3 continued

2. ~~Exhaust shall be taken from a point within 12 inches (305 mm) of the floor.~~

3. ~~The ventilation system shall be designed to operate at a negative pressure in relation to the surrounding area.~~

~~**5307.5.2 Emergency alarm system.** An emergency alarm system shall comply with all of the following:~~

1. ~~Continuous gas detection shall be provided to monitor areas where carbon dioxide can accumulate.~~

2. ~~The threshold for activation of an alarm shall not exceed 5,000 parts per million (9,000 mg/m³).~~

3. ~~Activation of the emergency alarm system shall initiate a local alarm within the room or area in which the system is installed.~~

5307.3.2 Gas detection system. Where ventilation is not provided in accordance with Section 5307.3.1, a gas detection system shall be provided in rooms or indoor areas and in below-grade outdoor locations with insulated carbon dioxide systems. Carbon dioxide sensors shall be provided within 12 inches (305 mm) of the floor in the area where the gas is expected to accumulate or other *approved* locations. The system shall be designed as follows.

1. Activate an audible and visible supervisory alarm at a normally attended location upon detection of a carbon dioxide concentration of 5,000 ppm (9000 mg/m³).

2. Activate an audible and visible alarm within the room or immediate area where the system is installed upon detection of a carbon dioxide concentration of 30,000 ppm (54 000 mg/m³).

CHANGE SIGNIFICANCE: Several revisions have occurred to the provisions for beverage dispensing systems utilizing liquid carbon dioxide.

The first change is in terminology. These systems are referred to as insulated liquid carbon dioxide systems. This term is consistent with the language in NFPA 55, *Compressed Gases and Cryogenic Fluids Code*, which is referenced in Section 5301.1 and therefore applies to the entire chapter. The CO_2 used at these facilities is a liquefied gas. It has been refrigerated to a temperature of -109°F where it turns to liquid. However, it is not a cryogenic fluid. To be classified as a cryogenic fluid, the liquid temperature must be below -130°F (-89.9°C).

The section requiring a permit is deleted. This is consistent with changes in Section 105.7 which removed the permit for liquid CO_2 systems in beverage dispensing and replaced it with the reference in Table 105.6.8 for a permit when the quantity of CO_2 exceeds 100 pounds.

Mechanical ventilation or gas detection is still required, however revisions occurred in the ventilation section to clarify that where ventilation is provided, it must be designed to ventilate all areas where released carbon dioxide could accumulate. Additionally, rather than listing the specific design criteria for the ventilation system, a reference to Section 5004.3 is inserted. Section 5004.3 contains the requirements previously found in the 2015 IFC Section 5307.5.1, plus additional requirements, such as providing a manual shutoff, prohibiting the recirculation of exhausted air, and continuous operation. This reference to Section 5004.3 now includes these additional safeguards.

The requirements for a gas detection system were also revised. Section 5307.3.2 specifies that the CO_2 sensors must be located within 12 inches of the floor. Now, two alarm levels are required. The lower threshold is at 5,000 ppm (OSHA Permissible Exposure Limit – 8-hour time weighted average). When this concentration is detected, an audible and visible supervisory alarm is activated at a normally attended location. This location would be an approved location at the facility, and one that is expected to be occupied at all times that the business is open. The second threshold is at 30,000 ppm (NIOSH Short Term Exposure Limit – 15-minute exposure). When this threshold is reached, an audible and visible alarm is required in the room or area where the insulated liquid carbon dioxide system is installed.

5307.4

Carbon Dioxide Enrichment Systems

International Code Council®

Carbon dioxide storage vessel used in a carbon dioxide enrichment system.

CHANGE TYPE: Addition

CHANGE SUMMARY: Carbon dioxide enrichment systems are now regulated by the *International Fire Code* when the system contains more than 100 pounds of CO_2, or when the refill connection is remote from the tank or vessel.

2018 CODE: **5307.4 Carbon dioxide enrichment systems.** The design, installation and maintenance of carbon dioxide enrichment systems with more than 100 pounds (45.4 kg) of carbon dioxide, and carbon dioxide enrichment systems with any quantity of carbon dioxide having a remote fill connection, shall comply with Sections 5307.4.1 through 5307.4.7.

5307.4.1 Documentation. The following information shall be provided with the application for permit:

1. Total aggregate quantity of liquid carbon dioxide in pounds or cubic feet at *normal temperature and pressure.*
2. Location and total volume of the room where the carbon dioxide enrichment operation will be conducted. Identify whether the room is at grade or below grade.
3. Location of containers relative to equipment, building openings and *means of egress.*
4. Manufacturer's specifications and pressure rating, including cut sheets, of all piping and tubing to be used.
5. A piping and instrumentation diagram that shows piping support and remote fill connections.
6. Details of container venting, including but not limited to vent line size, material and termination location.
7. Alarm and detection system and equipment, if applicable.
8. Seismic support for containers.

5307.4.2 Equipment. Pressure relief, vent piping, fill indicators, fill connections, vent terminations, piping systems and the storage, use and handling of the carbon dioxide shall be in accordance with Chapter 53 and NFPA 55.

5307.4.3 Gas detection system. A gas detection system complying with Section 916 shall be provided in rooms or indoor areas in which the carbon dioxide enrichment process is located, in rooms or indoor areas in which container systems are located, and in other areas carbon dioxide is expected to accumulate. Carbon dioxide sensors shall be provided within 12 inches (305 mm) of the floor in the area where the gas is expected to accumulate or leaks are most likely to occur. The system shall be designed as follows:

1. Activates a low-level alarm upon detection of a carbon dioxide concentration of 5,000 ppm (9000 mg/m³).

2. Activates a high-level alarm upon detection of a carbon dioxide concentration of 30,000 ppm (54 000 mg/m³).

5307.4.3.1 System activation. Activation of the low-level gas detection system alarm shall automatically:

1. Stop the flow of carbon dioxide to the piping system.
2. Activate the mechanical exhaust ventilation system.
3. Activate an audible and visible supervisory alarm signal at an *approved* location within the building.

Activation of the high-level gas detection system alarm shall automatically:

1. Stop the flow of carbon dioxide to the piping system.
2 Activate the mechanical exhaust ventilation system.
3. Activate an audible and visible evacuation alarm both inside and outside of the carbon dioxide enrichment area, and the area in which the carbon dioxide containers are located.

5307.4.4 Pressurization and ventilation. Rooms or indoor areas in which carbon dioxide enrichment is provided shall be maintained at a negative pressure in relation to the surrounding areas in the building. A mechanical ventilation system shall be provided in accordance with the *International Mechanical Code* that complies with all of the following:

1. Mechanical ventilation in the room or area shall be at a rate of not less than 1 cfm per square foot [0.00508 m³/(s • m²].
2. When activated by the gas detection system, the mechanical ventilation system shall remain on until manually reset.
3. The exhaust system intakes shall be taken from points within 12 inches (305 mm) of the floor.
4. The ventilation system shall discharge to the outdoors in an *approved* location.

5037.4.5 Signage. Hazard identification signs shall be posted at the entrance to the room and indoor areas where the carbon dioxide enrichment process is located, and at the entrance to the room or indoor area where the carbon dioxide containers are located. The sign shall be not less than 8 inches (200 mm) in width and 6 inches (150 mm) in height and indicate:

CAUTION—CARBON DIOXIDE GAS
VENTILATE THE AREA BEFORE ENTERING.
A HIGH CARBON DIOXIDE (CO_2) GAS CONCENTRATION
IN THIS AREA CAN CAUSE ASPHYXIATION.

5307.4.6 Seismic and structural design. Carbon dioxide system containers and piping shall comply with the seismic design requirements in Chapter 16 of the *International Building Code* and shall not exceed the floor loading limitation of the building.

5307.4 continues

5307.4 continued

5307.4.7 Container refilling. Carbon dioxide containers located indoors shall not be refilled unless filled from a remote connection located outdoors.

CHANGE SIGNIFICANCE: The proliferation of indoor marijuana cultivation facilities is expanding rapidly across the country. One of the processes utilized to stimulate plant growth uses a carbon dioxide enrichment system to increase the carbon dioxide concentration within the growing rooms. This process creates a potential asphyxiation hazard as the gas is intentionally being introduced into indoor occupiable rooms and areas.

These requirements apply when either the carbon dioxide system contains more than 100 pounds of CO_2, or when the refill connection for CO_2 is remote from the tanks or vessels. Note, even though the requirements may apply for a carbon dioxide enrichment system with remote fill connections, the requirement for an operational permit only applies when the quantity exceeds 100 pounds. Also, Section 5307.4.7 requires that CO_2 containers located inside a building must be refilled from a remote connection located outdoors.

A gas detection system is required in rooms or areas where carbon dioxide is utilized or stored. Carbon dioxide sensors shall be located within 12 inches of the floor and shall provide both a low-level and high-level alarm. The low-level alarm threshold is 5,000 ppm (OSHA Permissible Exposure Limit – 8-hour time weighted average). When the low-level alarm activates, the flow of carbon dioxide ceases, the mechanical exhaust system is started, and an audible and visible supervisory alarm occurs at an approved location. The high-level alarm threshold is 30,000 ppm (NIOSH Short Term Exposure Limit—15-minute exposure). When the high-level alarm activates, the flow of carbon dioxide ceases and the mechanical exhaust system is started, if these actions have not already occurred at the low-level alarm threshold. In addition, the high-level alarm requires an audible and visible evacuation alarm inside the monitored areas and outside of the carbon dioxide enrichment areas.

Mechanical ventilation is required for rooms where carbon dioxide enrichment is utilized. The normal ventilation for the room shall operate at a negative pressure in relation to the other areas of the building. When either the low-level alarm or high-level alarm activate, a mechanical exhaust system shall activate which exhausts 1 cfm per square foot of the room. The exhaust system must remain active until manually shut off, and discharge outdoors.

Signs are required outside of the rooms where carbon dioxide enrichment occurs. These warning signs are required at the entrances to these rooms to notify occupants of the potential hazard.

5707

Mobile Fueling Operations

CHANGE TYPE: Addition

CHANGE SUMMARY: On-demand mobile fueling is allowed to occur at approved locations and under the control of a permit issued by the fire code official.

2018 CODE: **5707.1 General.** On-demand mobile fueling operations that dispense Class I, II and III liquids into the fuel tanks of motor vehicles shall comply with Sections 5707.1 through 5707.6.3.

> **Exception:** Fueling from an *approved* portable container in cases of an emergency or for personal use.

5707.1.1 Approval required. Mobile fueling operations shall not be conducted without first obtaining a *permit* and approval from the *fire code official*. Mobile fueling operations shall occur only at *approved* locations.

5707.2 Mobile fueling vehicle. An on-demand mobile fueling vehicle shall be one of the following:

1. A vehicle that has chassis-mounted tanks or containers where the aggregate cargo capacity does not exceed 1200 gallons (4592 L). A mobile fueling vehicle with a mounted *tank* in excess of 110 gallons (415 L) shall comply with the requirements of Section 5706.6, Section 5707 and NFPA 385.
2. A vehicle that carries a maximum of 60 gallons (227 L) of motor fuel in metal safety cans *listed* in accordance with UL 30 or other *approved* metal containers, each not to exceed 5 gallons (19 L) in capacity. Containers shall be secured to the mobile fueling vehicle except when in use.

5707 continues

International Code Council®

This mobile fueling operation is occurring in a parking lot at a business office.

5707 continued

The mobile fueling vehicle shall comply with all local, state and federal requirements. The mobile fueling vehicle and its equipment shall be maintained in good repair.

5707.3 Required documents. Documents developed to comply with Sections 5707.3.1 through 5707.3.3 shall be updated as necessary by the *owner* of the mobile fueling operation and shall be maintained in compliance with Section 108.3.

5707.3.1 Safety and emergency response plan. Mobile fueling operators shall have an *approved* written safety and emergency response plan that establishes policies and procedures for fire safety, spill prevention and control, personnel training and compliance with other applicable requirements of this code.

5707.3.2 Training records. Mobile fueling vehicles shall be operated only by designated personnel who are trained on proper fueling procedures and the safety and emergency response plan. Training records of operators shall be maintained.

5707.3.3 Site plan. Where required by the *fire code official*, a site plan shall be developed for each location at which mobile fueling occurs. The site plan shall be in sufficient detail to indicate: all buildings, structures, *lot lines*, property lines and appurtenances on site and their use or function; all uses adjacent to the *lot lines* of the site; fueling locations; the location of all storm drain openings and adjacent waterways or wetlands; information regarding slope, natural drainage, curbing, impounding and how a spill will be kept on the site property; and the scale of the site plan.

5707.4 Mobile fueling areas. Mobile fueling shall not occur on public streets, *public ways* or inside *buildings*. Fueling on the roof level of parking structures or other *buildings* is prohibited.

5707.4.1 Separation. Mobile fueling shall not take place within 25 feet (7620 mm) of buildings, property lines or combustible storage.

Exception: The *fire code official* shall be authorized to decrease the separation distance for dispensing from metal safety cans or other *approved* metal containers in accordance with Section 5707.2.

Where dispensing operations occur within 15 feet (4572 mm) of a storm drain, an *approved* storm drain cover or an *approved* equivalent method that will prevent any fuel from reaching the drain shall be used.

5707.4.2 Sources of ignition. Smoking, open flames and other sources of ignition shall be prohibited within 25 feet (7620 mm) of fuel dispensing activities. Signs prohibiting smoking or open flames within 25 feet (7620 mm) of the vehicle or the point of fueling shall be prominently posted on the mobile fueling vehicle. The engines of vehicles being fueled shall be shut off during fueling.

5707.5 Equipment. Mobile fueling equipment shall comply with Sections 5707.5.1 through 5707.5.4.

5707.5.1 Dispensing hoses and nozzles. Where equipped, the dispensing hose shall not exceed 50 feet (15 240 mm) in length. The dispensing nozzles and hoses shall be of an *approved* and *listed* type.

5707.5.2 Fuel limit. Mobile fueling vehicles shall be equipped with a fuel limit switch set to a maximum of 30 gallons (116 L) and a nozzle or other *approved* device that, when activated, immediately causes flow of fuel from the mobile fueling vehicle to cease.

5707.5.3 Fire extinguisher. An *approved* portable fire extinguisher complying with Section 906 with a minimum rating of 40-B:C shall be provided on the mobile fueling vehicle with signage clearly indicating its location.

5707.5.4 Spill kit. Mobile fueling vehicles shall contain a minimum 5-gallon (19 L) spill kit of an *approved* type.

5707.6 Operations. Mobile fueling vehicles shall be constantly attended during fueling operations with brakes set and warning lights in operation. Mobile fueling vehicles shall not obstruct emergency vehicle access roads.

5707.6.1 Dispensing hose. Where equipped, mobile fueling vehicles shall be positioned in a manner to preclude traffic from driving over the dispensing hose. The dispensing hose shall be properly placed on an *approved* reel or in an *approved* compartment prior to moving the mobile fueling vehicle.

5707.6.2 Drip control. Operators shall place a drip pan or an absorbent pillow under the nozzle and each fuel fill opening prior to and during dispensing operations to catch drips.

5707.6.3 Spill reporting. Spills shall be reported in accordance with Section 5003.3.1.

CHANGE SIGNIFICANCE: Regulations for mobile fueling have been added to the code to address the process of mobile fueling. Mobile fueling is occurring in many locations and delivers the fuel to the location of the vehicle, rather the vehicle driving to the service station.

The mobile fueling operator is required to obtain a permit to conduct this operation. These requirements place the responsibility on the mobile fueling operator to take safety precautions at the site of fueling. As part of the permitting process, the fire code official reviews the proposed locations for fueling. Mobile fueling can only occur at locations which have been approved by the fire code official, but once the site is approved, fueling can occur at the site repeatedly.

This section provides a path for mobile fueling and provides the fire code official the explicit authority to regulate such operations.

Mobile fueling is an emerging industry that has had significant consumer demand both nationally and internationally. The practice is already occurring and these regulations provide for safe operations and consistent application.

6104.3

Location of LP-gas Containers

CHANGE TYPE: Modification

CHANGE SUMMARY: New Footnote g specifies separations between above-ground LP-gas containers and public ways.

2018 CODE:

TABLE 6104.3 Location of LP-Gas Containers

LP-Gas Container Capacity (water gallons)	Minimum Separation Between LP-gas Containers and Buildings, Public Ways[g] or Lot Lines of Adjoining Property that Can Be Built Upon		Minimum Separation Between LP-gas Containers[b, c] (feet)
	Mounded or underground LP-gas Containers[a] (feet)	Above-ground LP-gas Containers[b] (feet)	
Less than 125[c, d]	10	5[e]	None
125 to 250	10	10	None
251 to 500	10	10	3
501 to 2,000	10	25[e, f]	3
2,001 to 30,000	50	50	5
30,001 to 70,000	50	75	(0.25 of sum of diameters of adjacent LP-gas containers)
70,001 to 90,000	50	100	
90,001 to 120,000	50	125	

For SI: 1 foot = 304.8 mm, 1 gallon = 3.785 L.

a. Minimum distance for underground LP-gas containers shall be measured from the pressure relief device and the filling or liquid-level gauge vent connection at the container, except that all parts of an underground LP-gas container shall be not less than 10 feet from a building or lot line of adjoining property that can be built upon.

b. For other than installations in which the overhanging structure is 50 feet or more above the relief-valve discharge outlet. In applying the distance between buildings and ASME LP-gas containers with a water capacity of 125 gallons or more, not less than 50 percent of this horizontal distance shall also apply to all portions of the building that project more than 5 feet from the building wall and that are higher than the relief valve discharge outlet. This horizontal distance shall be measured from a point determined by projecting the outside edge of such overhanging structure vertically downward to grade or other level upon which the LP-gas container is installed. Distances to the building wall shall be not less than those prescribed in this table.

c. Where underground multicontainer installations are composed of individual LP-gas containers having a water capacity of 125 gallons or more, such containers shall be installed so as to provide access at their ends or sides to facilitate working with cranes or hoists.

d. At a consumer site, if the aggregate water capacity of a multicontainer installation, comprised of individual LP-gas containers having a water capacity of less than 125 gallons, is 500 gallons or more, the minimum distance shall comply with the appropriate portion of ~~Table 6104.3~~ this table, applying the aggregate capacity rather than the capacity per LP-gas container. If more than one such installation is made, each installation shall be separated from other installations by not less than 25 feet. Minimum distances between LP-gas containers need not be applied.

e. The following shall apply to above-ground containers installed alongside buildings:
   1. LP-gas containers of less than a 125-gallon water capacity are allowed ~~next to the building they serve~~ without a separation distance where in compliance with Items 2, 3 and 4.
   2. Department of Transportation (DOTn) specification LP-gas containers shall be located and installed so that the discharge from the container pressure relief device is not less than 3 feet horizontally from building openings below the level of such discharge and shall not be beneath buildings unless the space is well ventilated to the outside and is not enclosed for more than 50 percent of its perimeter. The discharge from LP-gas container pressure relief devices shall be located not less than 5 feet from exterior sources of ignition, openings into direct-vent (sealed combustion system) appliances or mechanical ventilation air intakes.
   3. ASME LP-gas containers of less than a 125-gallon water capacity shall be located and installed such that the discharge from pressure relief devices shall not terminate in or beneath buildings and shall be located not less than 5 feet horizontally from building openings below the level of such discharge and not less than 5 feet from exterior sources of ignition, openings into direct vent (sealed combustion system) appliances, or mechanical ventilation air intakes.
   4. The filling connection and the vent from liquid-level gauges on either DOTn or ASME LP-gas containers filled at the point of installation shall be not less than 10 feet from exterior sources of ignition, openings into direct vent (sealed combustion system) appliances or mechanical ventilation air intakes.

f. This distance is allowed to be reduced to not less than 10 feet for a single LP-gas container of 1,200-gallon water capacity or less, provided such container is not less than 25 feet from other LP-gas containers of more than 125-gallon water capacity.

g. Above-ground LP-gas containers with a water capacity of 2,000 gallons or less shall be separated from public ways by a distance of not less than 5 feet.

6104.3 continues

International Code Council®

Separation between this LP-gas container and the public way is based on whether the container is greater than 2,000 gallon water capacity.

6104.3 continued

Containers with a water capacity greater than 2,000 gallons shall be separated from public ways in accordance with this table.

CHANGE SIGNIFICANCE: The first of two changes, in Footnote e, removes the phrase "next to buildings" and addresses container separation in the same language as the remainder of the table.

The second change in this table creates modifications to the LP-gas container separation distances between containers and public ways. "Public way," as defined in Chapter 2 is a "street, alley or other parcel of land open to the outside air leading to a street, that has been deeded, dedicated or otherwise permanently appropriated to the public for public use and which as clear width and height of not less than 10 feet (3048 mm)."

Now, above-ground LP-gas containers with a water capacity of 2,000 gallons or less shall be separated from public ways by a distance of not less than 5 feet. Containers with a water capacity greater than 2,000 gallons shall be separated from public ways in accordance with Table 6104.3.

This proposal was approved as it more closely aligns Chapter 61 with NFPA 58, *Liquefied Petroleum Gas Code*, regarding the concept of public way. The modification provides a compromise that still uses the concept of public way but the separation requirement will now only apply to above-ground containers.

Additionally, pressure relief valves are provided for smaller containers and they must be located not less than 3 feet horizontally from building openings below the level of such discharge and shall not be beneath buildings unless the space is well ventilated to the outside and is not enclosed for more than 50 percent of its perimeter.

This change also correlates the IFC with industry terms to avoid confusion and misapplication of the code.

PART 6

Appendices A through N

- **Appendix A** Board of Appeals
  No changes addressed
- **Appendix B** Fire-flow Requirements for Buildings
  No changes addressed
- **Appendix C** Fire Hydrant Locations and Distribution
  No changes addressed
- **Appendix D** Fire Apparatus Access Roads
  No changes addressed
- **Appendix E** Hazard Categories
- **Appendix F** Hazard Ranking
  No changes addressed
- **Appendix G** Cryogenic Fluids—Weight and Volume Equivalents
  No changes addressed
- **Appendix H** Hazardous Materials Management Plan (HMMP) and Hazardous Materials Inventory Statements (HMIS) Instructions
  No changes addressed
- **Appendix I** Fire Protection Systems—Noncompliant Conditions
  No changes addressed
- **Appendix J** Building Information Sign
  No changes addressed
- **Appendix K** Construction Requirements for Existing Ambulatory Care Facilities
  No changes addressed
- **Appendix L** Requirements for Fire Fighter Air Replenishment Systems
  No changes addressed
- **Appendix M** High-Rise Buildings—Retroactive Automatic Sprinkler Requirements
  No changes addressed
- **Appendix N** Indoor Trade Shows and Exhibitions

Appendices A through N contain provisions for establishing boards of appeals, fire department access, fire-flow requirements for buildings, locations for fire hydrants, inspection criteria for noncompliant fire protection systems, building information signs, retroactive installation of fire sprinklers in existing high-rise buildings, and indoor trade shows and exhibits. They contain information concerning the classification of hazardous materials and the equivalent weights and volumes of the most commonly found cryogenic fluids. Appendices A through D and H through N do not have the force of law unless they are adopted by the jurisdiction. Appendices E through G are intended only as information sources that supplement the requirements of the *International Fire Code* (IFC) and are not intended to be adopted. ■

E102.1.7.1

Hazard Classification of Oxidizers

APPENDIX N

Indoor Trade Shows and Exhibitions

E102.1.7.1

Hazard Classification of Oxidizers

CHANGE TYPE: Modification

CHANGE SUMMARY: This change revised the oxidizer classification of sodium dichloro-s-triazinetrione anhydrous (sodium dichloroisocyanurate anhydrous).

2018 CODE: **E102.1.7.1 Examples of liquid and solid oxidizers according to hazard.**

Class 4: *No change to text…*

Class 3: ammonium dichromate, calcium hypochlorite (over 50 percent by weight), chloric acid (10 percent maximum concentration), hydrogen peroxide solutions (greater than 52 percent up to 91 percent), mono-(trichloro)-tetra-(monopotassium dichloro)-penta-s-triazinetrione, nitric acid, (fuming – more than 86 percent concentration), perchloric acid solutions (60 percent to 72 percent by weight), potassium bromate, potassium chlorate, potassium dichloro-s-triazinetrione (potassium dichloro-isocyanurate), potassium perchlorate (99 percent), potassium permanganate (greater than 97.5 percent), sodium bromate, sodium chlorate~~,~~ and sodium chlorite (over 40 percent by weight) ~~and sodium dichloro-s-triazinetrione anhydrous (sodium dichloro-isocyanurate anhydrous)~~.

Class 2: barium bromate, barium chlorate, barium hypochlorite, barium perchlorate, barium permanganate, 1-bromo-3-chloro-5, 5-dimethylhydantoin, calcium chlorate, calcium chlorite, calcium hypochlorite (50 percent or less by weight), calcium perchlorate, calcium permanganate, calcium peroxide (75 percent), chromium trioxide (chromic acid), copper chlorate, halane (1, 3-dichloro-5, 5-dimethylhydantoin), hydrogen peroxide (greater than 27.5 percent up to 52 percent), lead perchlorate, lithium chlorate, lithium hypochlorite (more than 39 percent available chlorine), lithium perchlorate, magnesium bromate, magnesium chlorate, magnesium perchlorate, mercurous chlorate, nitric acid (more than 40 percent but less than 86 percent), perchloric acid solutions (more than 50 percent but less than 60 percent), potassium peroxide, potassium superoxide, silver peroxide, sodium chlorite (40 percent or less by weight), sodium dichloro-s-triazinetrione anhydrous (sodium dichloroisocyanurate anhydrous), sodium perchlorate, sodium perchlorate monohydrate, sodium permanganate, sodium peroxide, sodium persulfate (99 percent), strontium chlorate, strontium perchlorate, thallium chlorate, urea hydrogen peroxide, zinc bromate, zinc chlorate and zinc permanganate.

Class 1: *No change to text…*

CHANGE SIGNIFICANCE: This modification reclassifies water treatment and swimming pool treatment products into a lesser hazard category among the four oxidizer classes, moving it from Class 3 to Class 2. Consequently, the maximum allowable quantity (MAQ) for solid water treatment chemicals increases from 10 to 250 lbs.

Sodium dichloro-s-triazinetrione anhydrous (sodium dichloroisocyanurate anhydrous) is a Class 2 oxidizer. The 2018 code reclassifies sodium dichloro-s-triazinetrione anhydrous (sodium dichloroisocyanurate anhydrous) as a Class 2 oxidizer which correlates with NFPA 400, *Hazardous Materials Code.*

A body of work has been completed to evaluate the tests and criteria for oxidizers and formulated products containing oxidizers. Objective testing was completed to evaluate the classification of oxidizers based on detailed measurements of burning rates and heat release measurements. This testing has confirmed that the correct classification of sodium dichloro-s-triazinetrione anhydrous (sodium dichloroisocyanurate anhydrous) is a Class 2 Oxidizer.

International Code Council®

Many pool chemicals with the primary constituent of sodium dichloro-s-triazinetrione anhydrous are reclassified as Oxidizer Class 2.

Chapter N

Indoor Trade Shows and Exhibitions

International Code Council®

Large trade shows and exhibits must comply with specific requirements in Appendix N, when the appendix is adopted.

CHANGE TYPE: Addition

CHANGE SUMMARY: This new chapter provides specific regulation to address the hazards associated with large trade shows and exhibitions. Many of the regulations can be found elsewhere in the IFC, but this appendix provides a single location with the regulations dealing with these events.

2018 CODE: Due to the extent of the code text, an outline of the appendix is shown below. For the complete text, see the 2018 *International Fire Code*.

APPENDIX N
INDOOR TRADE SHOWS AND EXHIBITIONS

N101 GENERAL

N101.1 Scope.

N101.1.1 Nonsprinklered buildings.

N101.1.2 Sprinklered buildings.

N101.2 Permit required.

N101.3 Application.

N102 DEFINITIONS

N102.1 Definitions.

N103 PUBLIC SAFETY FOR EVENTS

N103.1 Fire safety and evacuation plan.

N103.2 Fire watch personnel.

N103.3 Crowd managers.

N104 INTERIOR FINISH AND DECORATIVE MATERIALS

N104.1 General.

N104.2 Interior wall and ceiling finish.

N105 MULTIPLE-LEVEL BOOTHS

N105.1 Construction documents.

N105.2 Structural design.

N105.3 Means of egress.

N105.4 Automatic sprinkler systems.

N105.5 Inspection.

N105.6 Fire alarm and detection.

N106 COVERED BOOTHS

N106.1 Automatic sprinkler systems.

N106.2 Fire alarm and detection.

N107 DISPLAY AND STORAGE OF HAZARDOUS AND COMBUSTIBLE MATERIALS

N107.1 Hazardous materials.

N107.1.1 Display near exit.

N107.2 Storage of combustible materials.

N107.3 Vehicles.

N107.3.1 Batteries in vehicles.

N107.3.2 Vehicle fuel.

N107.3.2.1 Fueling within the structure.

N107.3.2.2 Vehicle fuel tanks.

N107.3.2.3 Vehicle fuel systems.

N107.3.2.4 Vehicle fuel tank openings.

N107.3.3 Obstruction by vehicles.

N107.3.4 Gas-powered vehicles.

N107.3.4.1 Shutoff valves.

N107.3.4.2 Battery hot lead.

N107.3.4.3 Dual-fuel vehicles equipped to operate on gasoline.

N107.3.5 Competitions or demonstrations.

N107.4 Fueled equipment other than vehicles.

N107.5 Liquid propane gas containers.

N107.5.1 LP-gas containers exceeding 12 pounds (5 kg) of water capacity.

N107.5.2 Where more than one LP-gas container is present in the same area.

N107.5.3 Equipment for LP-gas containers.

N107.5.4 Securing of LP-gas containers.

N107.5.5 Spare LP-gas containers.

N107.6 Cooking and open-flame devices.

N108 MEANS OF EGRESS

N108.1 Means of egress from the indoor trade show or exhibition area.

N108.2 Design of means of egress.

N108.3 Aisles and corridors.

N109 REFERENCED STANDARDS

CHANGE SIGNIFICANCE: Appendix N was created to address the hazards that are associated with larger, more complex trade shows and exhibitions. Although many of these requirements are already included in various locations in this code, some of the more important items, such as requirements for covered booths and multiple-level booths, are not. In an attempt to make it easier for those organizing exhibitions and for individual exhibitors who are unfamiliar with the fire code requirements, Appendix N has the requirements gathered in a central location along with references to other requirements located within the code. Section N101

Chapter N continues

Chapter N continued

limits the application of these requirements to larger trade shows. The appendix does not apply to trade shows or exhibitions where:

- The exhibit area is less than 1,500 square feet in a nonsprinklered building with at least two remote exits, or
- The exhibit area is less than 4,500 square feet in a sprinklered building with at least two remote exits.

However, in either of the above situations, if exhibit booths have more than one level, then the appendix applies. Section N103 requires a fire safety and evacuation plan, fire watch and crowd managers. These requirements can also be found in IFC Chapter 4.

Section N104 requires interior finish, decorations and furniture to comply with requirements in Chapter 8.

Section N105 contains new requirements specific to booths or exhibits that have more than one level. The booths must be designed by a registered design professional and meet the structural requirements in Chapter 16 of the *International Building Code*. Multiple-level booths with an occupant load greater than 10 on a single upper level must be provided with two means of egress. Multiple-level booths must be equipped with fire sprinklers when they exceed 400 square feet on any level, and equipped with a fire alarm system when they exceed 120 square feet on any level.

Section N106 requires a covered exhibit booth to be equipped with fire sprinklers when the booth exceeds 100 square feet on any level, and to be equipped with a fire alarm system when it exceeds 120 square feet on any level.

Additional requirements address use of LP-gas, cooking and other open flames, indoor display of vehicles, means of egress, and obstruction of egress paths.

Appendix N can be adopted by jurisdictions looking for specific regulations on this subject, or used as a guide where it is not. Note that the provisions contained in this appendix are not mandatory unless specifically referenced in the adopting ordinance.

INDEX

G

H

I

L

M

N